UNDERSTANDING THE UNIVERSE

THE IMPACT OF SPACE ASTRONOMY

Based on Talks given at the
UN/IAU International Seminar on the Occasion of UNISPACE 82,
Hofburg, Vienna, Austria, 12 August 1982

edited by

RICHARD M. WEST

General Secretary, International Astronomical Union

D. Reidel Publishing Company

A MEMBER OF THE KLUWER ACADEMIC PUBLISHERS GROUP

Dordrecht / Boston / Lancaster

Library of Congress Cataloging in Publication Data

Main entry under title:

Understanding the Universe.

   Bibliography: p.
   Includes index.
   1. Space astronomy. I. West, Richard M., 1941–
QB136.U47   1983   522'.6   83–13910
ISBN-13:978-94-009-7213-1   e-ISBN-13:978-94-009-7211-7
DOI: 10.1007/978-94-009-7211-7

Published by D. Reidel Publishing Company
P.O. Box 17, 3300 AA Dordrecht, Holland

Sold and distributed in the U.S.A. and Canada
by Kluwer Academic Publishers,
190 Old Derby Street, Hingham, MA 02043, U.S.A.

In all other countries, sold and distributed
by Kluwer Academic Publishers Group,
P.O. Box 322, 3300 AH Dordrecht, Holland

CONTENTS

CONTENTS

UNDERSTANDING THE UNIVERSE

THE IMPACT OF SPACE ASTRONOMY

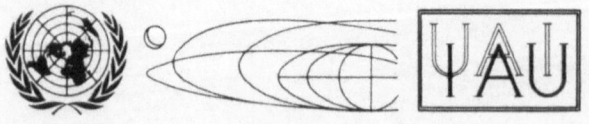

Editor's Preface

There is hardly any field of human endeavour which is more fundamental than the study of our surroundings . We have always wanted to learn what was behind our horizon , beyond the next mountain , on the other side of the ocean , on the next planet , at the end of the Universe .

We have come a long way since our early ancestors gazed upon the sky in amazement . Giant optical and radio telescopes now allow us to "see" the early epochs of the Universe , revealing phenomena beyond our comprehension . Spacecrafts with on-board astronomical instrumentation circle the Earth and fly to the limits of the Solar System , providing invaluable new information about nearby and distant objects .

Many people have the intuitive feeling that it is "easier and better" to study the Universe from above the Earth's atmosphere . However , this is only partially true in as much as electromagnetic radiation of certain wavelengths (e.g. X-rays) does not penetrate the atmosphere and can only be studied from balloons and spacecrafts . The advent of space-borne astronomy has not made ground-based observations obsolete - on the contrary , it is only thanks to the combination of the two that we have now a vastly more comprehensive picture of the Universe than just a few decades ago .

This book attempts to explain why and how this is so . It is based on lectures given by eminent scientists at the UN-IAU International Astronomy Seminar of UNISPACE 82 , in Vienna , Austria on 12 August 1983 and its five major chapters lead the reader from the nearby Sun to the most distant regions and the earliest times of the Universe . It is written at a level which is easily understood by the interested layman and centres on some of the most <u>fundamental</u> problems in modern astronomy and astrophysics . What has space - and ground-

R. M. West (ed.), Understanding the Universe, vii–viii.
© *1983 by D. Reidel Publishing Company.*

based astronomy told us so far and how can we best proceed ? What do we hope to learn during the next coming years ? (It was never the intention to include everything; certain subjects like the exploration of the planets have been extensively discussed elsewhere) .

I am greatly indebted to the authors for having so well transformed their talks into publishable articles . Without their friendly collaboration this book would never have appeared . Thanks are also due to the officials of UNISPACE 82 , in particular Professors Yash Pal and Alla Massevich and N. Jasentuliyana , as well as to my predecessor as IAU General Secretary , Professor P.A. Wayman , for support before , during and after the meeting. It is tragic to note the untimely death of the President of IAU , Professor M.K.V. Bappu on 19 August 1982 . Professor Bappu was looking forward to give the first talk in Vienna , but was unable to do so because of illness ; Professor J.-C. Pecker took his place at very short notice .

The camera-ready manuscript was typed by Ms. Patricia Smiley of the IAU Secretariat in Paris ; it is a pleasure to acknowledge her efficient and accurate work . I am also thankful to my collaborators in Munich , Mrs. E. Völk , Messrs. J. Leclercqz , C. Madsen , J. Quebatte and H. Zodet , for technical help .

I sincerely hope that this thought-provoking volume will contribute to spreading the knowledge about astronomy and some of its achievements , in particular among those who have not regularly followed recent developments . Then it will have served its purpose in the best possible way .

May 1983   **Richard M. West**
                                               General Secretary
                                    International Astronomical Union

Foreword by the President of the
International Astronomical Union

Man has always been fascinated by the stars . The starry skies have held him in wonder since ancient times and his curiosity and imagination , thus stimulated , have led to the accumulation of much folklore . Astronomy has therefore been the forerunner of the physical sciences . It has given us the sciences of mechanics , of optics , of spectrum analysis , and much of the progress of twentieth century physics has been closely interwoven with the skill we have displayed in utilizing the resources of this vast cosmic laboratory about us . And so , Astronomy is a science that is at the frontiers of human knowledge ; its pursuit continuously stimulates the considerable technological refinement that has been a characteristic of our times and which is inextricably linked with our daily living . The study of radiation originating from these diverse objects located in the depths of space , is the astronomer's principal tool and it is amazing how much of Nature's mysteries can be probed into by the power of the human mind and its capacity of reasoning . For mankind , the message of astronomy has a dichotomy ; it provides a feeling of humility at gauging Man's place in the Universe and a sensation of pride in his ability to unravel so much of its hidden mystery by the power of his intellect .

Munich , 1 August 1982   [†]M. K. V. Bappu

Foreword by the Secretary General of United Nations
Conference on the Exploration and Peaceful Uses of Outer Space

In the process of organising the Second United Nations Conference on the Exploration and Peaceful Uses of Outer Space (UNISPACE 82) , I felt it would be most useful to build around this Conference a number of other events and activities that might contribute to creating a greater awareness about the role and potential of Space , and at the same time enhance the involvement of both scientists/technologists and the lay public (including especially , young people) . Accordingly , we had organised - with the help of various countries and organisations - exhibitions , poster and essay contests , demonstrations , seminars etc. Some of these were held before the Conference and others during UNISPACE 82 in Vienna itself . Among the latter was the International Astronomy Seminar sponsored jointly by the International Astronomical Union (IAU) and the United Nations .

This Seminar was held in Vienna on 12 August 1982 and covered a number of topics . Eminent astronomers from many countries attended , making the presentations and discussions very stimulating and interesting . It was indeed unfortunate that Professor M.K.V. Bappu (the President of the International Astronomical Union) , who was to present a paper on "The Sun and Sunlike Stars" could not attend as he was seriously ill , and has since expired . As one of the chief organisers of this seminar , its success was a tribute to him .

Astronomy is the oldest "space science" - and yet , the advent of space technology has led to a feeling of liberation even in the field of astronomy . New discoveries to date are impressive , especially in areas of ultraviolet , X-ray and gamma-ray astronomies , but much exciting work lies ahead . The launching of large telescopes , improvement of sensors and data handling systems and the possibility of astronomers everywhere in the world using these magnificent robots , with data-links , again provided through space technology , will revolutionise the field of astronomy .

R. M. West (ed.), Understanding the Universe, xi–xii.
© *1983 by D. Reidel Publishing Company.*

One of the excellent traditions of astronomical research is the very positive spirit of international cooperation , even in today's difficult political environment . The International Astronomical Union - which includes astronomers from the US , USSR , Europe and also from Asia , Africa , Australia and Latin America - is one symbol of such cooperation . I hope the UNISPACE Conference has also made a significant contribution to this atmosphere of working together in this field and other areas of space activity .

However , such cooperation between scientists may not for long be insulated from the pulls and pressures of the overall international situation . In space itself , we now have the beginnings of an extremely costly and tremendously dangerous arms race . This can only further aggravate the dangers of a cataclysmic war - one that will see the end of civilization as we know it and make almost permanently uninhabitable vast areas of the world . In a few years it will be too late ; the time to stop this desecration of outer space is now .

In this task , scientists have a bigger role to play than others , for vested interests often seek to confuse the layman with scientific-sounding , jargonised rationales for "defensive" weapons in space . In this regard , UNISPACE has urged all countries to adhere to the Outer Space treaty - in both letter and spirit - and has called on " all nations , in particular those with major space capabilities...to contribute actively to the goal of preventing an arms race in outer space and to refrain from any action contrary to that aim " .

I am delighted that the IAU and the United Nations have arranged to publish the proceedings of the International Astronomy Seminar . This will be of great interest to astronomers , other scientists and educated public the world over . I hope it will also serve to remind them not only of the benefits of international cooperation , but also its fragility . There is need to create a climate in which international cooperation can continue to thrive , and astronomers - along with other scientists - have special responsibility in this .

June 1983   **Yash Pal**

CHAPTER I

L'ATMOSPHERE DU SOLEIL ET DES ETOILES *
The Atmosphere of the Sun and the Stars

by Jean-Claude PECKER

Institut d'Astrophysique ,
98 , bis blvd Arago , 75014 , Paris ,
France

Summary

The Sun is a very pleasant star to study: it is the closest , by far , and we can study all the details of its surface .   Actually , the number of photons reaching the Earth from the Sun is tremendously bigger than the number of photons from the rest of the Universe , meaning that we have much more information from our star Sun than from the rest of the Universe .

In the recent years , the progress made in the study of the various solar regions has greatly benefitted from the extension to the UV , XUV , and IR parts of the spectrum , as a result of the blooming of space research .

---

*   This lecture was given at rather short notice , in place of the lecture which was prepared on the same subject by Professor Vainu Bappu , and because of his bad health . A few days later , we all learnt of the untimely death of our friend Vainu Bappu , President of the International Astronomical Union . In spite of the lack of originality of this talk , which has been using many of the slides prepared for it by Vainu , and which has been , in many respects , a reiteration of ideas given at some recent occasions , the author wishes to dedicate this paper to the memory of a fine scientist , and of an old friend , to the memory of Vainu Bappu , whose personal contribution has been so important in the field covered (as seen hereafter) .   (The lecture was delivered in French , slides being commented in English) .

1

R. M. West (ed.), Understanding the Universe, 1–35.
© *1983 by D. Reidel Publishing Company.*

In the visible light , the information gathered allows us to better know the layers of the Sun's photosphere; only in the center of some intense and opaque spectral lines can we detect the effect of outer chromospheric layers; the visible corona is seen only during total eclipses , and to a rather limited extent , through the coronagraph . The XUV , the X-rays - and the radiowaves on the other side of the spectrum - have given us a good knowledge of the coronal regions . In particular , the spectra of strongly ionized ions (such as Fe XV , Fe XVI or even Fe XXIV) are good indicators of the very hot regions (several millions of degrees K) . The study in X-rays has shown the importance of coronal holes , often extended along solar meridian lines , from pole to pole . They seem to be associated with regions that are geomagnetically active . The study of the corona is a good basis for the study of the solar wind .

What concerns the stars , of course , we are not so well advanced ! But satellites such as Einstein have shown the evidence of the stellar coronae . One can almost say that all stars have coronae . This brings about a difficult problem , as no theory can yet explain this in a satisfactory way; they do not explain quantitatively either the coronal temperatures , or the distribution with photospheric temperature , or spectral type , of the coronal line intensities .

The UV measurements (by the satellites Copernicus and IUE) have shown that the stellar outer layers , affected by strong winds , by strong departures from equilibrium , cannot be described by the too few parameters previously used to model the stellar atmospheres: the physics is following these discoveries , and we now start to better understand the nature of phenomena in these outer layers .

One phenomenon essential in the understanding of stellar chromospheres is the so-called **Wilson-Bappu effect** . Its broad validity indicates some relation between the stellar luminosity and the turbulence: the structure of magnetic features is probably of paramount importance in accounting for the observed relations . The fine-structure of the Wilson-Bappu relation (effects of other parameters) should allow to understand this complex physical problem in a better way .

The Sun , and the stars , are **active** . One knows now the stellar activity cycles , comparable to the 11-year periodicity in the activity of the Sun . The theory of the cycle is still in infancy: but , as it now stands , it shows that it can be understood only by taking into account the coupling between rotation , convective motions and magnetism . The fact that the internal structure of stars (and of the Sun) is badly known is certainly a heavy handicap . But quick progress is made due to the fact that two important groups of observations give almost direct information on what happens in the deep solar interior: one is the observations of the flux of **neutrinos** (an observation which is impossible to extend to the stellar case) and the study of solar **seismology** (which can be extended to stars in a foreseeable future) .   The latter has developed quickly and should benefit , in the coming years , from long series of unbiased observations , pursued on-board space stations .   Various types of oscillations and waves (pressure waves , gravity waves) affect the solar atmosphere and their theory gives much information about the structure of the convective layers of the internal parts of the Sun .

For us , the Sun , our star , is a unique laboratory .   But , if the stellar studies are much behind , one might expect some important discoveries in the coming years .   In the meantime , phenomena linked with "solar" - like magnetic activity   - are known in the case of many stars , and more will certainly be found in the coming years .   The reasons (resolving power in angle , time and spectral resolution) why the Sun has had for years a prominent place in stellar research will still be valid for a long time; but , more and more , stellar studies will be able to compete , and to extend our knowledge to the behaviour of atmospheres differing from the solar one by some essential parameters .   In a way , after a period when it was often said , as a sort of programme for further research , "the Sun is a star" or even "the active Sun is a star" , one could say now that "stars are suns" , implying the importance of stellar studies as a substitute for impossible experimentation on the Sun .   Solar and stellar physicists are indeed more and more speaking the same language .

1. Introduction

C'est une tradition: toute étude sur les atmosphères stellaires commence par l'analyse des méthodes appliquées à l'exploration de l'atmosphère solaire.

Le soleil est en effet l'étoile la plus proche de nous ; et de ce fait, nous sommes capables d'explorer tous les détails de sa surface, y compris, pendant les éclipses, dans les régions de la couronne. Le flux des photons solaires est énorme, si bien que l'on peut aussi étudier le spectre des phénomènes solaires avec un excellent pouvoir de résolution spectral. Enfin, on peut suivre l'évolution des phénomènes solaires avec une grande finesse, la constante de temps étant, pour les études solaires, extrêmement faible.

Depuis le milieu du siècle, l'extension des domaines explorés a permis de connaître encore mieux la physique solaire: ce fut d'abord l'ouverture du spectre radioélectrique, grâce à la radioastronomie. Les développements successifs des techniques spatiales ont permis ensuite l'ouverture des domaines de l'infrarouge, celle de l'ultraviolet, celle des rayons X, celle des rayons gamma: tout le spectre électromagnétique du Soleil nous est aujourd'hui connu - et, là aussi, la résolution spatiale, spectrale, temporelle, certainement encore perfectible, reste remarquable.

La résolution spectrale est un atout particulièrement utile: en effet elle permet la mesure du champ magnétique, celle des vitesses, en chaque point de la surface solaire ; on peut maintenant dresser des cartes très précises, très fines, du champ magnétique solaire (de ses diverses composantes), et des vitesses projetées le long de la ligne de visée, au moins dans les régions de la photosphère.

Ajoutons que du Soleil proviennent des flots de particules ; celles qui sont ionisées sont canalisées par les champs magnétiques terrestres ; elles constituent une partie importante du rayonnement cosmique ; les particules neutres, comme les neutrinos, sont à peine observables, mais la neutrino-astronomie du Soleil est déjà une discipline riche en enseignements de grande valeur.

Pendant longtemps au contraire (les quelques progrès actuels de l'astronomie des tavelures , pour spectaculaires et prometteurs qu'ils soient , s'ils changent nos perspectives , n'ont encore apporté des données utilisables que pour deux ou trois étoiles) on n'observera de la plupart des étoiles que la lumière totale , sans pouvoir séparer par l'observation directe les régions "actives" des autres régions , dites "calmes" , sans pouvoir séparer directement la couronne des régions plus froides de la chromosphère ; toutes les informations viendront d'une étude du spectre , avec une résolution spectrale modérée , avec une résolution temporelle non moins modérée .

Il est donc clair que le progrès essentiel peut venir de l'ouverture du spectre grâce à la recherche spatiale: d'ores et déjà les satellites UV comme Copernicus , ou IUE , les satellites X comme Uhuru et Einstein , les satellites gamma , comme SAS-2 ou COS-B - bien d'autres études encore , par ballons , fusées , satellites , en URSS , aux USA , en Europe , au Japon - ont permis , par l'exploration du spectre , l'exploration des couches différentes de l'étoile , responsables de ces divers rayonnements .

Nous examinerons d'abord quelques-uns des progrès accomplis dans la connaissance du Soleil grâce à la recherche spatiale , puis dans celle des étoiles . Ensuite , nous passerons en revue les progrès effectués (du "sol" terrestre) dans le domaine de la recherche solaire , au cours des récentes années ; et nous tenterons de prévoir comment des progrès comparables pourront être accomplis dans les cas des étoiles grâce à l'exploration spatiale bien sûr , mais aussi grâce à la patience des observateurs au sol .

La figure 1 , librement adaptée des documents publiés par la NASA , met en évidence cette impuissance essentielle des études stellaires: au manque de résolution spatiale , il faudra évidemment répondre . La réponse viendra , pour les étoiles les plus proches (et , parmi elles , les étoiles de plus grand diamètre) , de l'astronomie interférométrique à tavelures ("speckle interferometry") . Pour le plus grand nombre (en restant cependant dans le voisinage solaire) , c'est en augmentant le temps d'observation , et en étendant le domaine spectral exploré que l'on pourra procéder , indirectement , à l'étude de la physique des étoiles .

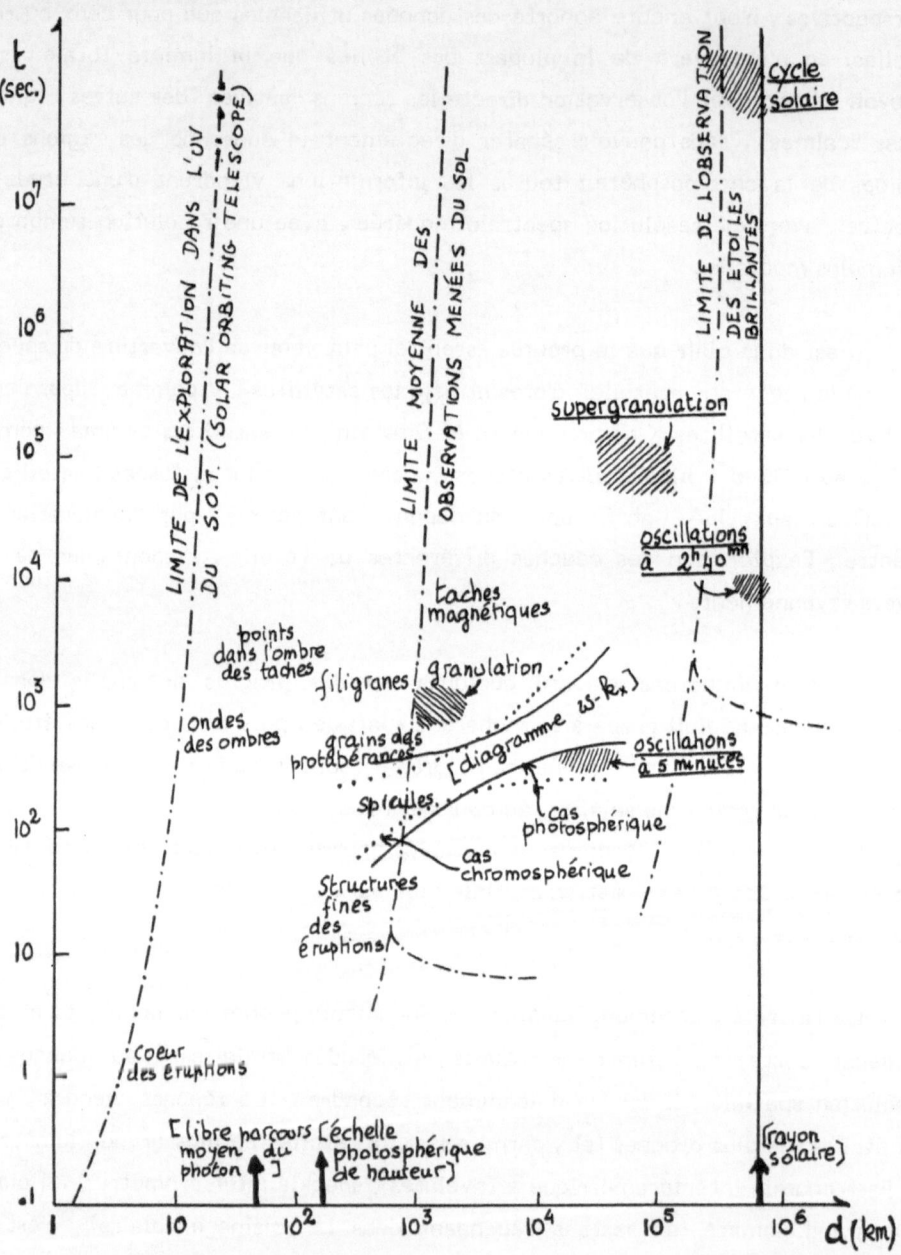

Figure 1   Diagram of solar phenomena (after original NASA documents)

In the abscissa , the logarithm of the typical size of solar structures . In the ordinate , the logarithm of the typical time-scale t , of solar structures . One can note the classical ω - k_x diagram located in the t - d system of coordinates , at its location , as commonly drawn . The dot-dashed lines show the limits to the left beyond which information cannot be obtained , in three cases ; from left to right: NASA S.O.T. in the UV ; ground-based solar observatories ; ground-based stellar observations for the brightest and closest stars . Three vertical arrows at the bottom indicate , from left to right , the mean free path of photons in the photosphere , the photospheric scale-height , and the solar radius , respectively .

---

2 . Le Soleil , des Rayons X aux Ondes Kilométriques

Dans le domaine visible , le spectre solaire , la lumière solaire provient essentiellement des régions les plus profondes de l'atmosphère d'où peuvent provenir directement des photons (régions les plus transparentes) - celles de la **photosphère** .

Cependant , le spectre est affecté , notablement au centre des raies intenses comme celles du calcium ionisé (raies H ou K) ou comme les raies de Balmer (Raie H α de l'hydrogène) , par les conditions physiques dans la **chromosphère** . Un grand nombre de raies moins intenses (par exemple les raies de résonance des métaux - mais seulement celles-là) sont affectées par des écarts à l'équilibre thermo-dynamique , qui se manifestent dès les régions extérieures de la photosphère. L'étude de ces régions est possible du sol , en permanence .

Du sol , ce n'est au contraire (et encore depuis les années 30 de notre siècle) que grâce au coronographe que l'on peut observer la **couronne** au bord du disque , en dehors des quelques minutes par an que durent , au mieux , les éclipses totales de Soleil , observées des rares points du globe où se produit telle ou telle d'entre elles .

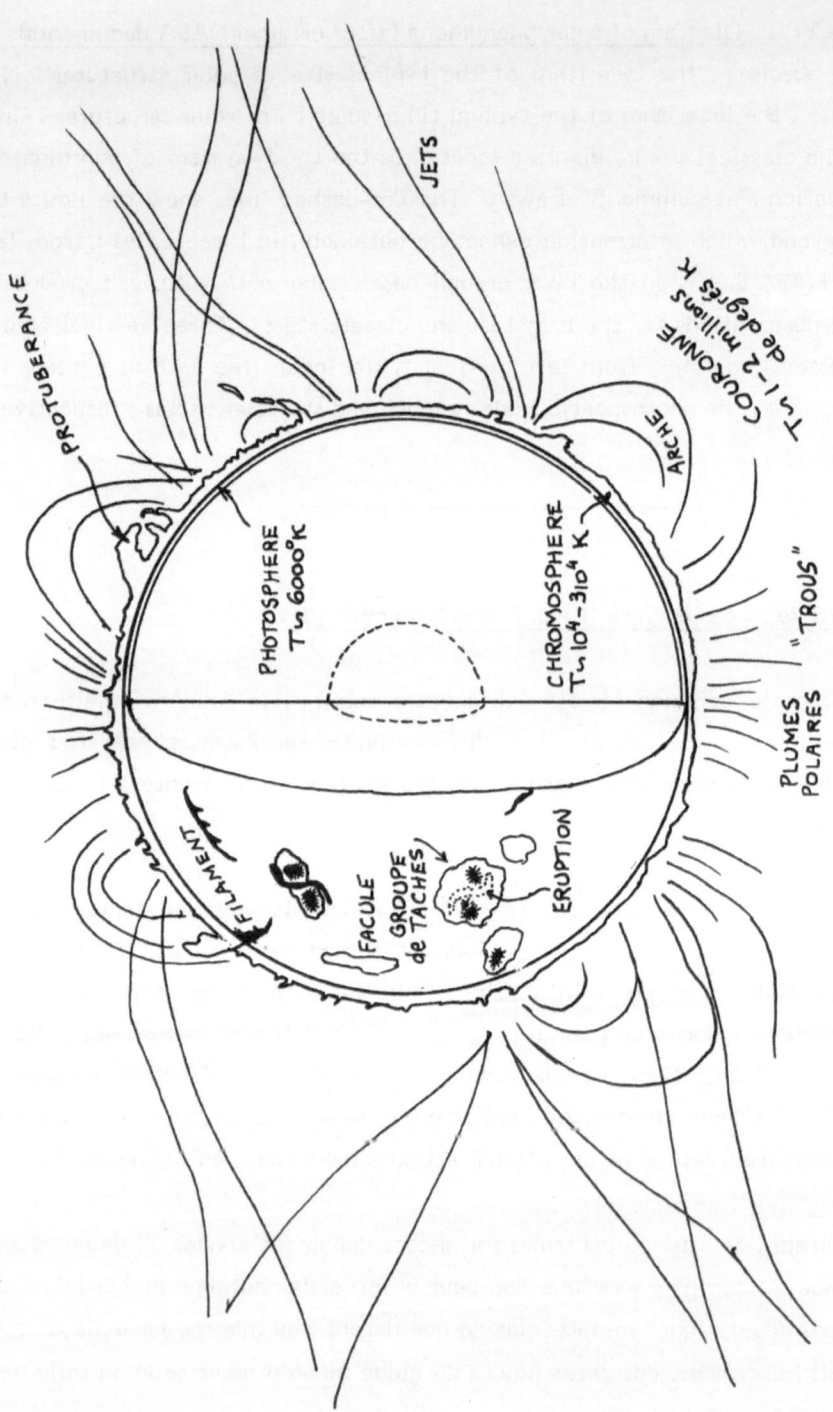

Les rayonnements du proche ultraviolet et de l'infrarouge , auxquels les couches photosphériques sont opaques , permettent d'explorer les régions proches du minimum de température , entre photosphère et chromosphère . Ballons et fusées , dès les années 50 , ont ainsi permis d'établir le premier modèle de ces régions , le BCA (Bilderberg continuous atmosphere) .

Mais surtout , l'exploration récente de l'UV lointain et du domaine X a fait apparaître de nombreuses raies en émission , dues à la présence d'ions de potentiel d'ionisation très élevé , couvrant un domaine très étendu , depuis les indicateurs de température chromosphérique , bien entendu (dans le proche UV , tout particulièrement le doublet de résonance de Mg II et les raies de Lyman de l'hydrogène neutre) ceux issus de la région de transition entre chromosphère et couronne (pratiquement non étudiée auparavant) grâce aux raies UV de C IV , Si I ,

------

Figure 2   The nomenclature of solar phenomena (schematical)

In the photosphere , the active phenomena are mainly faculae in the heart of which appear the smaller magnetic spots , with their umbra , and penumbra (facules , taches , ombre , pénombre) . In the chromosphere , active and often violent flares (éruptions) are observable , and have a short life-time compared with the life-time (several months) of an active area (région active) . Granules are small scale photospheric features while spicules are small scale chromospheric features ; but neither granular nor spicular structures have been shown in this drawing , as one can see on Figure 1 their caracteristic time and size. Prominences are cooler than the chromosphere and develop above it and they can be seen at the limb as the chromosphere in the $H\alpha$ Hydrogen line , being either quiescent or active , or even eruptive , and on the disk , as dark filaments above the chromospheric layers. The corona , whose structure is clearer on the X-ray pictures , such as indicated in Figure 3 , is very hot and its structure in jets , arches and loops is rather complex ; near the poles , there are often coronal holes (trous coronaux) and coronal polar plumes .

ou de O VI , jusqu'à ceux de couronne moyenne - vers un million de degrés - (comme les raies X de Si XI , Mg X ou Mg IX), et celles enfin indiquant la présence de régions extrêmement chaudes dans la couronne - plusieurs millions de degrés (Fe XV , Fe XVI , jusqu'à Fe XXIV...) . On notera que , contrairement aux observations coronographiques, c'est sur l'ensemble de la couronne (devant le disque donc , également) que portent désormais les observations .

Simultanément , les progrès de la radioastronomie , principalement grâce à l'emploi d'instrumentations à grand pouvoir angulaire de résolution , ont permis de compléter les informations venues de l'espace .

Si bien que l'étude des régions extérieures de l'atmosphère solaire a pu permettre d'en connaître la structure thermique , d'y déterminer la distribution des densités , d'y connaître les champs de vitesse , macroscopique et microscopique (voire "mésoscopique") , et même d'y mesurer les champs magnétiques , au moins dans leurs zones les plus basses . Les régions actives sont mieux connues que naguère ; la connaissance du **vent solaire** , et de ses caprices , a fait en peu d'années de grands progrès .

Nous nous limiterons à un petit nombre parmi les découvertes importantes encore récentes .

Tout d'abord à celle des **trous coronaux** .

Observée dans une bande spectrale telle que la bande 2-32 $\overset{o}{A}$, qui contient plusieurs raies de haute ionisation , la couronne montre une structure très hétérogène . Apparaissent comme des "trous" des régions étendues, de faible éclat , et dont l'étude montre qu'ils correspondent à des régions de très faible densité . Ces "trous" sont liés aux courants à très grande vitesse du vent solaire ; ils sont plutôt associés aux régions assez actives qu'aux régions calmes de la surface solaire . La structure des trous est souvent allongée le long d'un méridien et ils vont parfois d'un pôle à l'autre , comme s'ils n'étaient pas affectés par la rotation différentielle (un phénomène déjà constaté dans l'étude des phénomènes géomagnétiques , sans doute associés à la structure de la couronne) . Leur

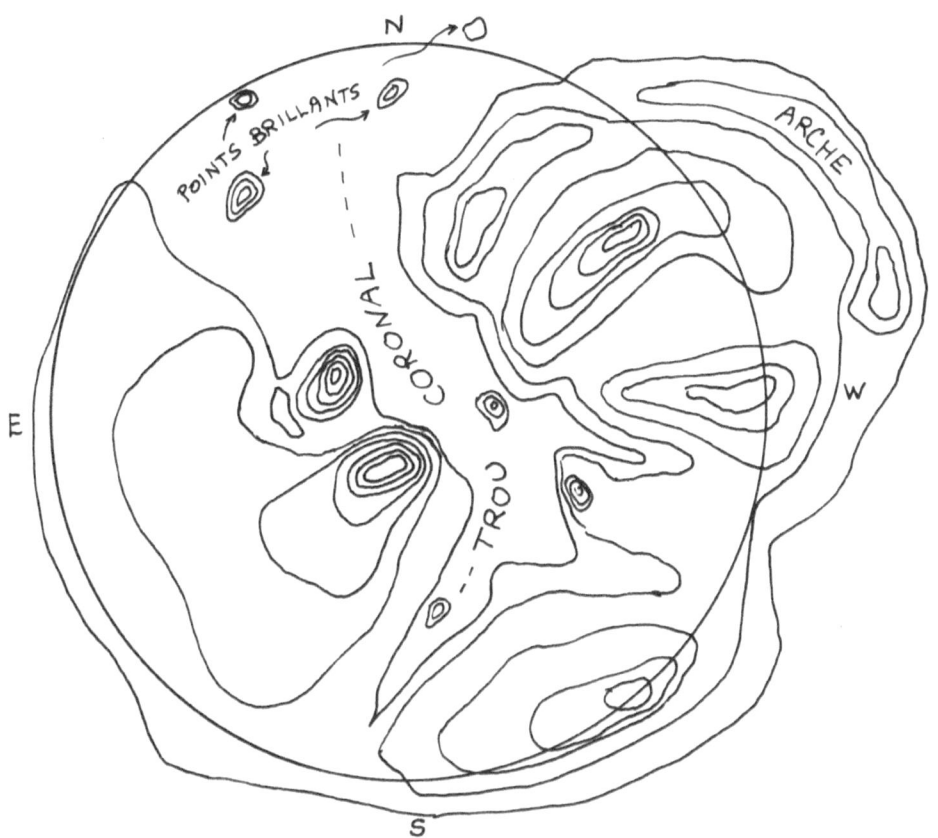

Figure 3   Coronal holes

This isophotal map of the corona, as observed in a line of a highly ionized
element, has been drawn by the author from some original NASA pictures, as
representative of what can be observed. One sees the large coronal hole,
extending across the equator, almost along a meridian circle. We note the
appearance of the corona above the limb, as it can be seen during an eclipse.
One may also remark a few coronal bright dots, visible not only in the hole but
also in the main corona.

température est sans doute un peu plus élevée que celle de la couronne moyenne ; on peut aussi les observer , dans leurs régions les plus basses , grâce à l'interférométrie dans les ondes centimétriques , à pouvoir de résolution angulaire élevé .

La mesure de la vitesse du **vent solaire** , dans les régions coronales , et principalement dans les trous coronaux , a permis d'obtenir des valeurs précises de 10 à 20 km s^{-1} , à comparer avec les valeurs de l'ordre de 50 km s^{-1} , observables à 1 <u>ua</u> de distance du Soleil . Ces valeurs sont d'ailleurs fonction de la phase du cycle solaire , et affectées par une dispersion naturelle due à l'inhomogénéité des mécanismes de perte de masse , et des conditions physiques dans les régions commandant , et affectées par , le vent solaire . De telles déterminations , faites grâce à la mesure des décalages spectraux des raies d'émission du domaine X , permettent de donner une base assez ferme aux théories du vent solaire .

Un autre aspect des découvertes dues à l'utilisation des engins satellisés est la précision apportée désormais dans la structure des **régions de transition** entre chromosphère et couronne . Le comportement des régions **actives** est assez différent de celui des régions **calmes** . Dans les régions actives , il semble bien que la dispersion des vitesses (ascendantes ou descendantes) des régions émissives soit plus grande , très nettement , que dans les régions calmes ; les mouvements ascendants semblent dominer dans les régions actives . Les vitesses non thermiques , à petite échelle , semblent passer par un maximum entre chromosphère et couronne ; leur décroissance vers l'extérieur est plus marquée pour les noeuds ascendants ou descendants des régions actives , que pour les parties moins mobiles des régions calmes ou actives . La détermination de la densité fait apparaître des densités dix fois plus élevées au-dessus des régions actives que des régions calmes , et une altitude en moyenne plus basse au-dessus de la photosphère . Ces données , encore peu étudiées , doivent permettre de mieux comprendre les mécanismes de conversion d'énergie mécanique en énergie magnétique .

En résumé - et sans donner d'autres exemples , car la littérature est devenue considérable - , il reste clair que c'est la connaissance des **mécanismes** physiques dans la chromosphère , dans la couronne active et dans le vent solaire , qui ont bénéficié le plus des progrès de la recherche spatiale . Qu'en est-il dans le domaine stellaire ?

3 . L'Etude des Etoiles dans les Domaines X et UV

Dire que le Soleil est une étoile prototype , c'est une banalité (sur laquelle nous reviendrons dans notre conclusion) . La connaissance du Soleil dans les domaines XUV a donc poussé les chercheurs à étudier le domaine stellaire XUV des spectres stellaires . Guidé par des théories très générales (et dépassées aujourd'hui) sur l'origine du chauffage de la couronne et de la chromosphère solaire , supposée associée à une zone convective épaisse , on chercha donc l'émission X dans certaines étoiles plutôt que dans d'autres .

D'emblée , les nouvelles observations mirent en évidence l'impossibilité de prévoir quelles étoiles auraient un rayonnement X élevé , et quelles en auraient un faible . Ainsi , dans l'étoile double **Alpha du Centaure** , c'est la composante K qui a une couronne brillante , et la composante G une couronne faible . On peut dire que toutes les étoiles , de tous les types , ont des **couronnes** , et qui se traduisent par des émissions X . Cette constatation est évidente depuis le satellite Einstein (Uhuru avait permis d'étudier les sources les plus brillantes du ciel - supernovae , QSS , etc. - mais peu d'étoiles normales) . Les flux observés s'étendent sur un grand domaine de variation , de 10^{26} à 10^{34} ergs par seconde . Cette mesure de la luminosité dans le domaine X montre aussi l'absence d'une corrélation évidente entre cette luminosité X et les propriétés de l'étoile (température effective , gravité) déduites des classifications spectrales basées sur les observations visuelles . Une corrélation semble en revanche être possible avec la **rotation** et le **magnétisme stellaire** . On notera en particulier que le flux X des étoiles jeunes O et B , aussi bien que celui des étoiles K et M , est mille à un million de fois plus élevé qu'il n'est prévu dans les théories simplistes antérieures à l'astronomie X des étoiles .

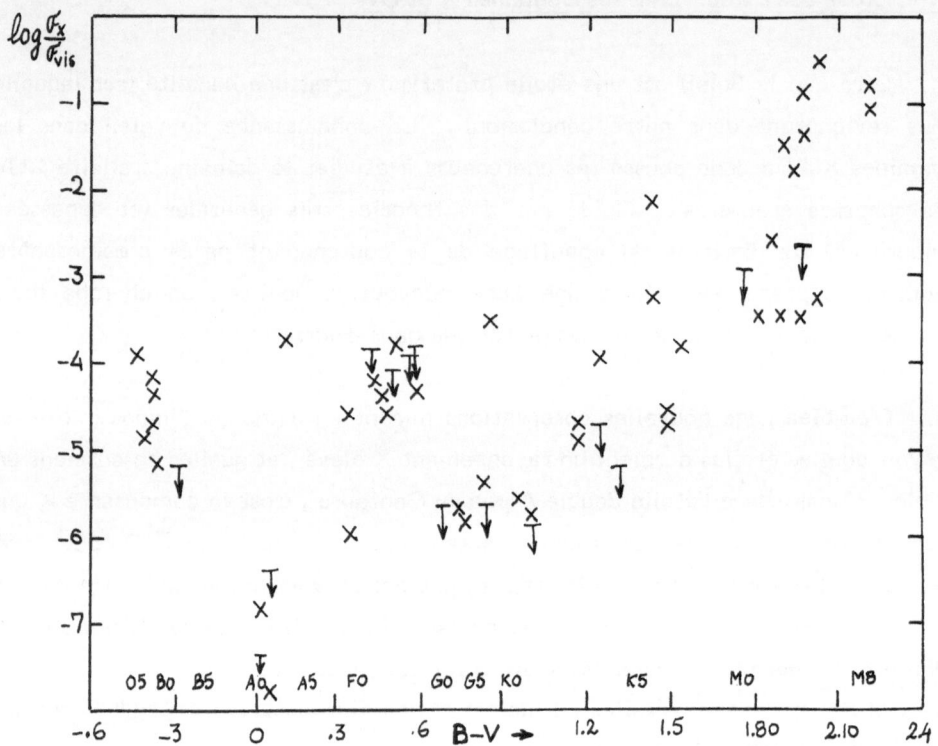

Figure 4   <u>Stellar X-ray fluxes</u>

In the abscissa , the color index (the associated main sequence spectral types are indicated) ; in the ordinate , the logarithm of the ratio of the flux in the X-ray region to the flux in the visible . The diagram is only for main-sequence stars . We notice the large natural scatter , indicating the need for additional parameters to describe stellar coronae , in addition to the effective temperature of the star and its gravity (the figure is drawn following various authors) .

On doit mentionner aussi, dans ce domaine, l'observation d'éruptions brutales de rayonnement X. Des étoiles comme YC CMi, ou BY Dra, ont montré ces éruptions X, simultanées avec les éruptions radio, et observables dans les raies d'émission visibles. Dans le second cas, les éruptions X semblent superposées à des variations cycliques d'activité X. Des sources comme Sgr A - le centre de la Galaxie - sont des sources X bien étudiées, également marquées par d'importantes éruptions.

Les observations XUV - Copernicus, ou IUE - ont permis la mesure du spectre d'un très grand nombre d'étoiles de tous les types, et en particulier des raies d'émission correspondant aux températures de la chromosphère. Comme dans le cas des émissions X, on s'est aperçu que le phénomène chromosphérique était très général. De la mesure des profils de raies stellaires, on peut bien

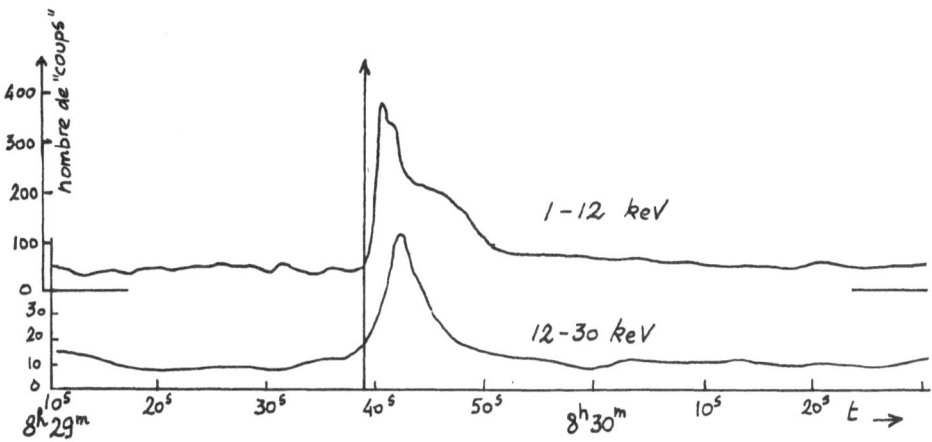

Figure 5   A flare in the radiosource Sgr A

One sees, in the course of a few seconds, the development of a flare at the galactic center (drawn by the author from the original results of Matsuoka and associates).

entendu déduire la distribution des vitesses: la forme (de type "P Cygni") de
nombreux profils de raies est l'indicateur d'un **vent stellaire** très important . Le
taux de perte de masse par vent stellaire semble plus élevé pour les étoiles de plus
forte luminosité ; l'existence d'un vent implique une décroissance assez lente de la
densité vers l'extérieur , et reste donc naturellement associée à des couronnes
étendues ; mais le rôle du champ magnétique dans la distribution de la perte de
masse avec la direction , et en général dans les phénomènes coronaux , est un
paramètre supplémentaire dont l'intervention atténue les corrélations claires
entre les deux phénomènes , couronne et vent . Bien entendu , les vents sont
variables , comme le sont , de façon générale , chromosphère et couronne , même
lorsque l'éclat total de l'étoile est constant , ou presque .

L'existence de vents affectant les étoiles jeunes , non encore arrivées
(T Tauri) ou arrivées depuis peu (étoiles O et B) , sur la série principale , pose un
problème . La théorie a priori de la contraction des masses gazeuses sphériques
semble en effet indiquer clairement que le temps de chute libre des couches
extérieures est grand ; pour ces étoiles jeunes , la nébuleuse protostellaire , ses
régions extérieures tout au moins , sont encore en train de tomber . Comment
concilier ceci avec l'existence de vents vers l'extérieur ?   Probablement en
admettant que les régions équatoriales et polaires ne jouent pas le même rôle ;
rotations assez rapides , champs magnétiques peut-être importants: le jeu des
forces gravitationnelles , inertielles , électromagnétiques , peut faire qu'il y a
encore "chute" à des latitudes stellaires élevées , et déjà "vent" au voisinage des
zones équatoriales...Mais le problème est posé ; et l'idée que nous venons
d'émettre n'en est qu'au stade d'une hypothèse de travail .

D'ores et déjà , la littérature consacrée aux chromosphères et aux couronnes
stellaires est donc imposante .

4 . Un Phénomène Chromosphérique: L'Effet Wilson-Bappu

Parmi les travaux qui ont prolongé vers les étoiles des phénomènes
chromosphériques bien connus sur le Soleil , figure la découverte , par **Olin Wilson**

et **Vainu Bappu** en 1957 , de l'effet qui porte leur nom et qui marqua une date importante dans la progression de nos idées sur les couches extérieures , non chromosphériques , des étoiles .

On sait que la raie K du calcium ionisé (observable du sol , et souvent aussi observée depuis l'espace - par exemple par le satellite OSO-8) est un indicateur excellent de l'activité solaire ; les images de la surface solaire faites dans la raie K mettent en évidence les régions actives (facules) , et le profil de la raie K est un excellent indicateur global de cette activité: le centre de la raie (K3) est sensible à l'activité , les ailes (K1) en sont peu affectées ; la séparation $\Delta\lambda_2$ des pics K2 , leur importance , sont de bons indicateurs des champs de vitesse , du gradient des propriétés physiques dans les régions hautes de la chromosphère solaire . L'évolution de la raie K entre les parties calmes et actives du Soleil est très marquée , et très significative . Elle a permis récemment de montrer , aux frontières de la région active , l'existence de zones turbulentes , résultant , dans ces régions très limitées , d'une conversion d'énergie magnétique en énergie mécanique .

En moyenne , la raie K du Soleil , qui varie avec la phase du cycle solaire (voir ci-après , paragraphe 5) , reste assez bien définie , notamment en ce qui concerne la séparation $\Delta\lambda_2$ des deux pics K2 .

Le phénomène découvert par Wilson et Bappu (ci-après désignés par les initiales WB) , c'est la variation de cette séparation avec la luminosité ($\Delta\lambda_2 \sim L^6$) , identique pour presque tous les types spectraux pour lesquels elle est mesurable. Ce phénomène est indicatif du rôle dominant d'un seul paramètre , la luminosité , comme essentiellement responsable de l'importance d'une seule autre quantité , la **turbulence** dans les zones chromosphériques . Nous noterons que cette relation ne saurait être vérifiée qu'en moyenne , compte tenu de ce qui est observé pour le Soleil , des régions calmes aux régions actives . Mais elle est cependant remarquablement bien établie , et la relation de WB est suivie avec une grande précision sur un grand intervalle de luminosité , des étoiles de la série principale aux supergéantes . La théorie de cette loi est encore fort peu avancée . Il est clair en tout cas que la séparation $\Delta\lambda_2$ des pics est indépendante

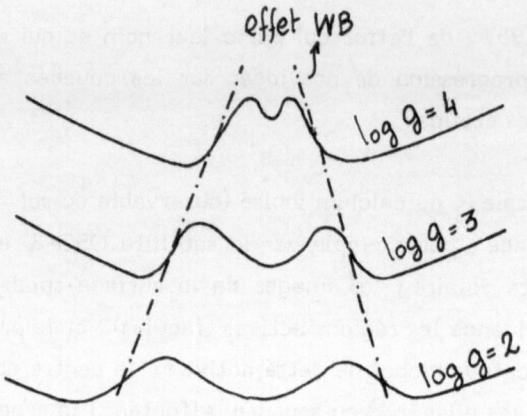

effet WB

$\log g = 4$

$\log g = 3$

$\log g = 2$

effet de la gravité
(et de la luminosité)   → λ

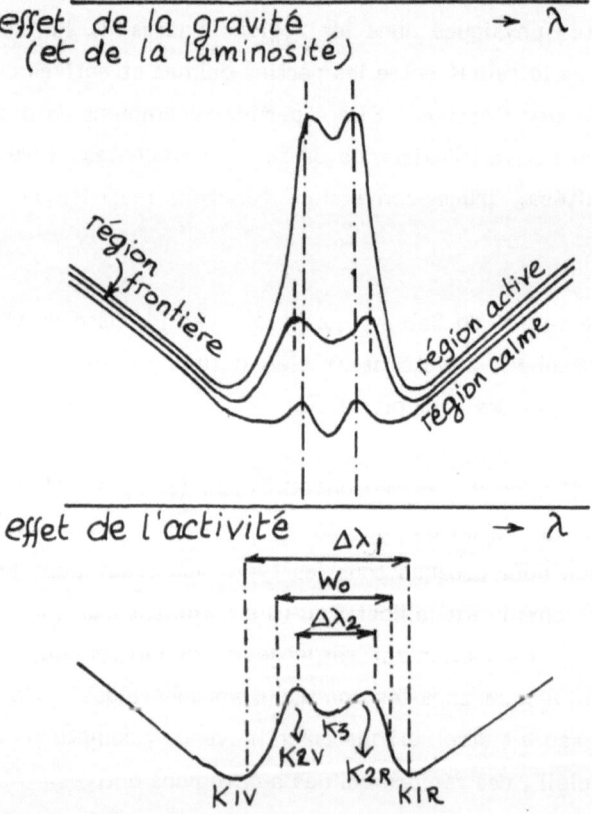

région
frontière

région active

région calme

effet de l'activité   → λ

$\Delta\lambda_1$

W_0

$\Delta\lambda_2$

K_3

K_{2V} K_{2R}

K_{1V} K_{1R}

définitions   → λ

Figure 6   <u>The behaviour of the K emission as an activity index (schematical)</u>

At the bottom , the definition of the various symbols characterizing the central part of the K line is given . In the middle , the differential behaviour of the K line from a quiet part of the Sun to the heart of a facula is shown. One should note the increase of $\Delta\lambda_2$ in the intermediate frontier region , when going from quiet to active areas .

At the top , for a given effective temperature , the behaviour of the K line with respect to gravity is described . We see that W_0 decreases a lot for the higher values of the gravity ; when studied in the HR diagram , one can see that the more important parameter is indeed the luminosity , to which W_0 is linked by the law $W_0 \sim L^6$, established by Wilson and Bappu .

---

de l'intensité I3 de K3: sans doute l'une de ces quantités (I3) dépend-elle des propriétés individuelles des tubes de force magnétiques , l'autre ($\Delta\lambda_2$) résultant essentiellement du nombre des tubes de force , chacun d'eux jouant le rôle de facteur essentiel de l'existence d'une région active .

La recherche de la structure fine de la relation de WB est une tâche importante , à laquelle l'étude (par des moyens spatiaux) de raies comparables à la raie K du Ca II , comme les raies h et k du Mg II , dans l'ultraviolet proche (comparables , mais plus sensibles aux phénomènes de la haute chromosphère) devrait apporter des arguments décisifs à la compréhension des chromosphères stellaires . De telles recherches ont été effectivement abordées à l'aide du satellite IUE .

Mais d'autres domaines essentiels de la recherche solaire sont en plein développement ; l'extension de ces travaux est-elle possible vers les étoiles ? Et d'autre part , la recherche spatiale peut-elle contribuer à la progression des connaissances dans tous les domaines , comme elle le peut dans les domaines étudiés ci-dessus ? Nous allons maintenant nous poser ces deux questions à propos de trois domaines importants de la recherche solaire d'aujourd'hui .

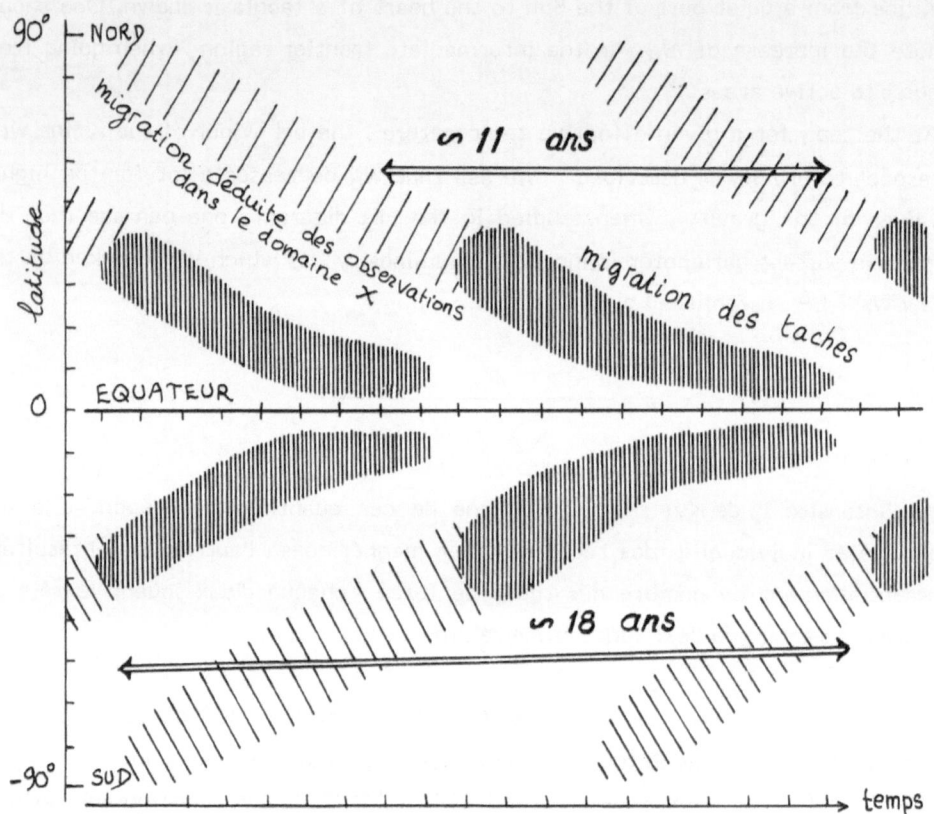

Figure 7   The Spörer butterfly diagram (schematical)

In the ordinate, the heliographic latitude, in the abscissa, the time. The presence of sunspots is shown schematically by a dark hatched area and the migration of the regions where sunspots appear towards the equator is obvious. At latitudes larger than 40° or 45°, one cannot observe spots, but bright points observed in X-rays, and other features strongly suggest the existence of an active region, which migrates as well, in agreement with the spot migration, as indicated (light hatched areas) in the figure.

5 . L'Activité Solaire

Un domaine de recherches qui , quoiqu'aussi ancien que la recherche solaire elle-même , continue à se développer , et se voit même dans une ère de nouvelle jeunesse , c'est l'étude de l'activité solaire , et tout particulièrement du phénomène général de l'activité solaire **cyclique** - et pas seulement de ses épiphénomènes (taches , éruptions...) qui accompagnent le développement global de l'activité .

Il est clair que cette question est étroitement liée à l'étude du magnétisme , et à celle de la rotation et de la rotation différentielle (effet différentiel selon la latitude , et selon la profondeur) .

Dans le domaine solaire , les études spatiales poursuivies dans le domaine X , et tout particulièrement l'étude des "points brillants" observables , très localisés, situés dans les régions des trous coronaux , ont permis de prolonger jusqu'aux pôles les lois que l'étude menée dans le domaine visible avaient permis de mettre en évidence . Ainsi la **"loi de Spörer"** indique-t-elle la baisse vers l'équateur , au cours d'un cycle de onze ans , de la latitude moyenne des taches: c'est le fameux "diagramme papillon" qui met le mieux en évidence ce fait bien connu .

Mais le diagramme papillon ne permet pas de savoir ce qui se passe à des latitudes supérieures à 45° . Aujourd'hui , et grâce aux points brillants X , on peut le prolonger jusqu'à des altitudes très élevées . On voit alors que les phénomènes peuvent être décrits plutôt comme une **migration** des zones actives des pôles vers l'équateur ; cette migration dure **18 à 20 ans** et se superpose à la migration la précédant et à celle la suivant: la "période" classique de **11 ans** correspond , en fait , à l'intervalle entre deux "migrations" successives ; et cette superposition des migrations successives a quelque peu tendance , en affectant par exemple le nombre R de taches de la surface totale , ou le géomagnétisme (qui est corrélé assez bien avec R , et qui dépend de la totalité de la surface solaire) , à masquer le phénomène important , cette migration de quelques 20 ans de durée .

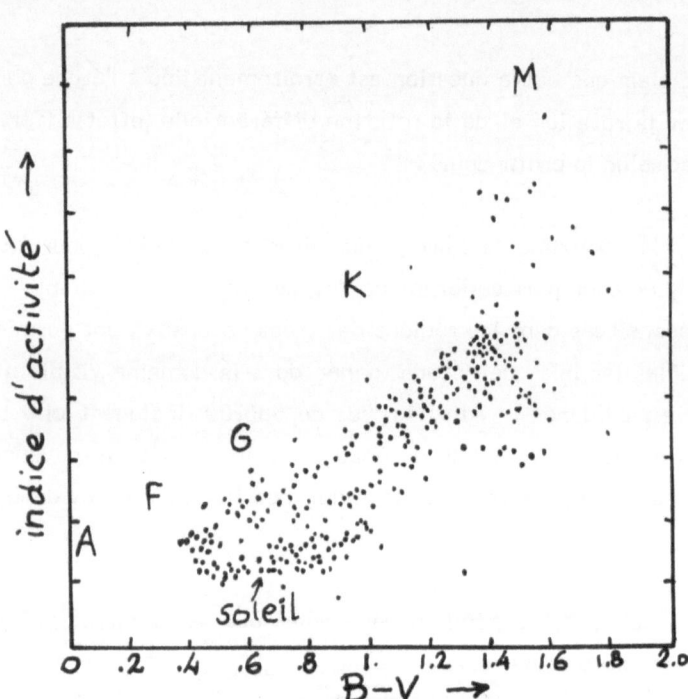

Figure 8   The stellar activity

An activity index is defined from the emission part of the K line ; its logarithm is in the ordinate while the color-index (B-V) is shown in the abscissa ; the spectral types , and the location of the Sun (a relatively quiet star) are shown. One can note the strong increase of activity towards the cooler stars of this diagram . (after Vaughan , A. ; Preston , G.W. ; 1980 , P.A.S.P. , 92 , p. 385)

La théorie du cycle d'activité a fait d'énormes progrès grâce au développement des **théories "dynamo"** , traitées aujourd'hui de façon non linéaire , et qui permettent , notamment , de prévoir une telle migration . Toutefois , la theorie dynamo rencontre encore de sérieuses difficultés , notamment en ce qui concerne la rotation différentielle du Soleil . De plus , de nombreuses étoiles sont des "rotateurs obliques": comme le montre l'analyse de leur spectre , l'axe de rotation et l'axe magnétique sont souvent très différents (dans le cas du Soleil , rotateur lent , ces deux axes sont très proches - 1° environ , si l'on en croit les statistiques portant sur une dizaine de cycles) . Mais s'ils ne sont pas confondus , alors la théorie dynamo ne saurait s'appliquer stricto sensu ; et il faudrait expliquer cette obliquité ; l'étude des étoiles magnétiques , l'application de la théorie à ces étoiles , devraient nous aider .

Toutefois , jusqu'à ces dernières années , on n'étudiait le magnétisme stellaire que sur des périodes comparables à leur période de rotation (quelques jours...) , grâce à l'effet Zeeman affectant les raies bien mesurables du spectre visible. Aujourd'hui , le problème a radicalement changé de dimensions grâce à la découverte , par Wilson et ses collaborateurs , de nombreuses étoiles , observées dans la raie K du Ca II (très sensible , comme dans le Soleil , aux importantes manifestations de l'activité stellaire dans les régions chromosphériques) , et qui se trouvent être affectées de variations cycliques d'activité . Les données permettent de trouver des comportements très différents: tantôt la variation à longue période (de l'ordre de quelques années) est simple , non perturbée ; tantôt elle est affectée par un "bruit" important , effet probable de fluctuations de l'activité à plus petite période , ou peut-être d'une distribution très localisée , sur la surface stellaire en rotation , des centres actifs . Dans d'autres cas encore , il semble que l'activité ne soit variable qu'à petite échelle de temps (comparable à la période de rotation) ; ou même qu'elle ne soit pas variable du tout . Il semble que l'activité stellaire soit nettement plus marquée pour les étoiles les plus froides de l'intervalle étudié .

Cette découverte majeure doit inciter les chercheurs utilisant des engins spatiaux , à suivre les mêmes étoiles pendant un long laps de temps , dans les raies de l'UV ou du domaine X , hypersensibles aux manifestations chromosphériques et

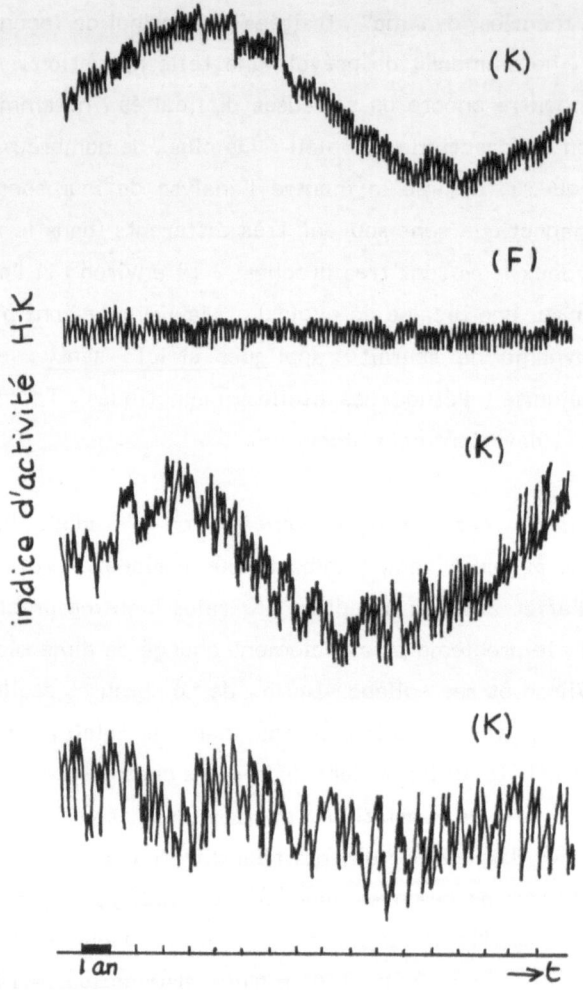

Figure 9   Fluctuations of stellar activity (schematical)

In the abscissa the time (one interval corresponds to one year); along the ordinate, the activity index. One sees different types of behaviour, which can be found in the stars surveyed by Wilson and his colleagues. This diagram has been drawn from published data by the author, and is representative (although far from realistic) of the different types of observations made so far (schematically from O.C. Wilson, 1978, Ap.J. p. 379)

coronales de l'activité stellaire . Elle doit aussi orienter les théoriciens des intérieurs stellaires , vers l'extension des solutions dynamos à de nombreux cas stellaires , voire leur modification , afin de rendre compte des rotateurs obliques . Mais , même sans une telle modification , elle doivent rendre compte , au moins dans leurs grandes lignes , des phénomènes qui , d'une étoile à l'autre , font varier la pseudo-période du cycle , ou la distribution , plus ou moins localisée , des centres actifs sur l'astre en rotation . C'est un domaine important de recherches qui devrait se développer rapidement .

6 . La Sismologie Solaire

Depuis des années , l'on connaît les étoiles pulsantes . Mais pendant longtemps , tout fut ignoré des "oscillations" solaires , des "pulsations" solaires .

La première observation qui en fut faite (par Leighton , Noyes et Simon) remonte à 1960 ; peu de temps après , grâce à des techniques différentes , elle fut confirmée , par Evans et Michard: de petites portions de la surface solaire oscillent avec une période de **5 minutes** environ ; ces oscillations affectent la photosphère . L'amplitude de la modulation des vitesses est de 0.5 km s^{-1} ; dans la photosphère , la vitesse de propagation des ondes est de l'ordre de 6 km s^{-1} . On s'aperçut assez vite que des surfaces assez importantes de la surface solaire oscillaient en phase . Et l'étude de ces ondes , de ces oscillations , déclencha bientôt une fructueuse compétition entre les observateurs , qui cherchèrent à mettre en évidence la structure fine des oscillations , à les décomposer en modes , cependant que les théoriciens travaillaient à l'étude , dans le Soleil , des instabilités vibrationnelles , et de la propagation des ondes .

Une sphère gazeuse peut osciller selon divers "modes" , comme un tuyau sonore , et on peut les classer en utilisant , comme pour les atomes , des nombres caractéristiques , analogues aux nombres quantiques . Ces modes sont de deux sortes: tout d'abord des modes g - lorsque les forces de gravitation constituent les forces de rappel - ces "ondes de gravité" sont en oeuvre , même dans un milieu incompressible (océans) ; puis les modes p , lorsque la force d'Archimède est la

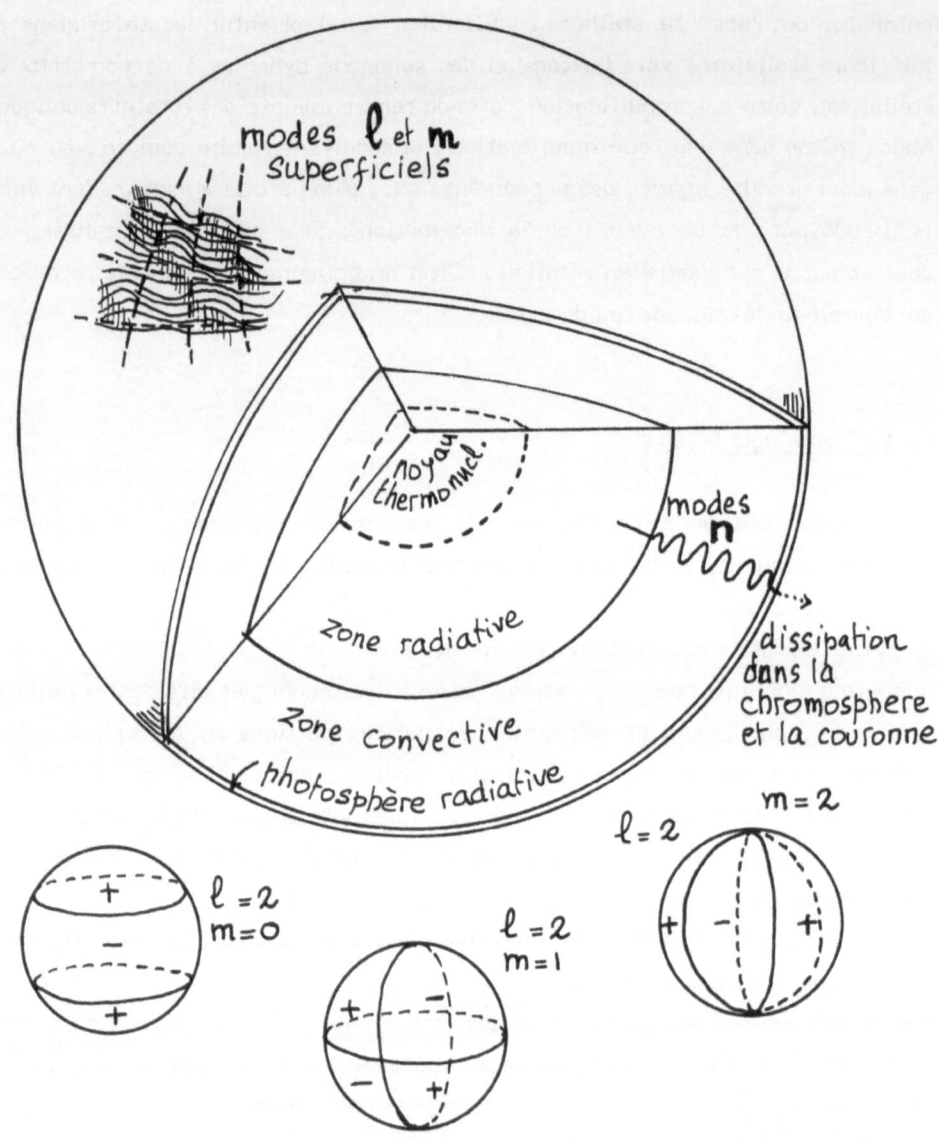

Figure 10   The oscillations of the solar sphere (schematical)

One can distinguish the radial modes , which are trapped in the convection alone ;
and the non-radial modes , defined by the number ℓ of nodal lines , and their
mutual arrangements (defined by the number m , as indicated in the three small
diagrams , for ℓ =2) . (schematically from Fossat , in ref 3)

force de rappel ("ondes de pression", comme celles en oeuvre dans un tuyau d'orgue). La théorie représente ces ondes avec trois "nombres", le nombre n, caractéristique des oscillations radiales, le nombre ℓ caractérisant, à la surface, le nombre de lignes nodales et le nombre m, caractérisant, pour une valeur donnée de ℓ, les dispositions mutuelles des lignes nodales.

Les observateurs cherchèrent d'abord comment se distribue l'énergie dans un diagramme traditionnel, le diagramme ω - k, où l'on porte en abscisse la fréquence spatiale k (horizontale, mesurée sur la surface solaire), et en ordonnée la fréquence temporelle ω; ce diagramme est reproduit, à petite échelle, sur la Figure 11. Dans ce diagramme, les modes p et les modes g sont localisés dans deux régions distinctes; entre ces régions, on trouve seulement des ondes évanescentes. Deubner, par l'étude d'une portion limitée de la surface solaire, montre qu'on peut alors améliorer le pouvoir de résolution temporel, et l'analyse (par des techniques utilisant la transformation de Fourier des signaux) aboutit à un diagramme dans le plan k - ω, où des modes p (de n = 0, le fondamental, à n = 12 environ) sont mis en évidence. Fossat et Grec, opérant sur le disque complet, et sur des périodes de temps très étendues (plus de cent heures, de façon continue, à partir du Pôle Sud) détectent principalement les modes radiaux de grand n (15 à 30) et, pour chacun, les modes correspondant aux petites valeurs de ℓ (1 à 4) - alors que seules les grandes valeurs de ℓ étaient accessibles aux études portant sur des portions trop petites de la surface solaire.

Sans entrer dans le détail de la théorie (menée d'abord par Ando et Osaki), on peut dire que les oscillations à cinq minutes sont des ondes stationnaires sonores (modes p) formées dans la "cavité sonore" existant entre le sommet de la zone convective d'ionisation de l'hydrogène, et la base de la chromosphère. Une autre cavité résonante chromosphérique, plus élevée, existe d'ailleurs, à l'origine des oscillations à 3 minutes détectées dans la chromosphère par l'étude des oscillations des raies de Lyman, menée dans l'ultraviolet (observations conduites sur le satellite OSO-8).

Un autre type d'ondes, dont la théorie est moins avancée (il peut s'agir, ce n'est pas sûr, d'ondes de gravité) a été détecté (essentiellement par des mesures

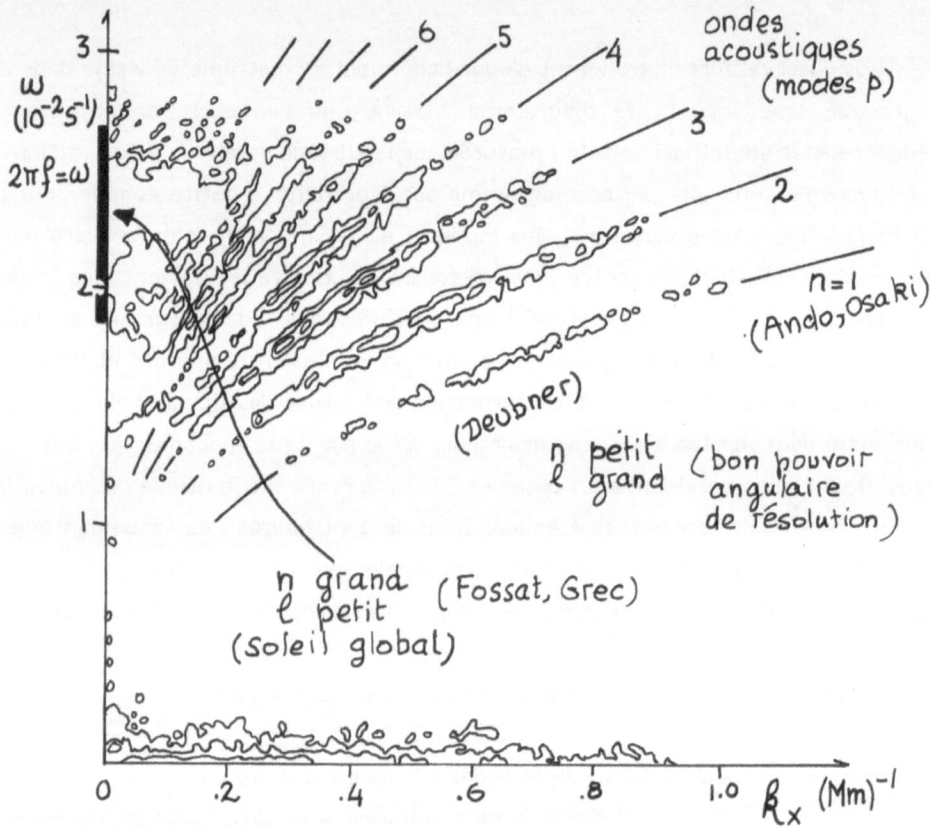

Figure 11   The ω - k_x power diagram

This diagram combines Deubner's original data , and Ando and Osaki's theoretical
predictions , which allow us to identify the radial modes in the power-diagram.
The regions of the diagram which have been investigated , using data concerning
the global Sun , from Fossat and Grec , have been indicated.   Original
Ando-Osaki theoretical results do not fit the Deubner observational data as well ;
but they could do so at the expense of some modification of the internal
structure: we have assumed the fit to be , as it can be , quite good .

indirectes du diamètre solaire) par Severny , Kotov et leurs collaborateurs , puis par d'autres groupes , avec une période de 2h 40,01 mn . L'amplitude est faible (quelques m s^{-1}) . L'étude de longues séquences d'observations implique , semble-t-il , la mise en évidence de divers harmoniques des ondes de gravité .

Il est clair que la sismologie solaire devrait à brève échéance (satellite DISCO) bénéficier grandement de possibilités d'observation continue , pendant les centaines ou les milliers d'heures consécutives qu'autorisent les observations menées depuis l'extérieur de l'atmosphère terrestre .

Dans le domaine stellaire , l'étude , au cours des décennies écoulées , de nombreuses étoiles pulsantes (ainsi les étoiles à courtes périodes dont β Cephei , ou δ Scuti , sont représentatives) fournit essentiellement le mode d'oscillation fondamental , et son (ses) premier(s) harmonique(s) . Il serait utile , évidemment , d'en connaître plus . Surtout , hors de la bande d'instabilité vibrationnelle qui localise sur le diagramme HR les étoiles pulsantes , il serait utile de connaître les oscillations , comparables à celles du Soleil , qui pourraient être détectées et analysées ; le fait qu'on tire les caractéristiques des oscillations de l'étude de la lumière totale du Soleil , avec une excellente résolution spectrale , mais sans avoir besoin de résolution angulaire , donne à cet égard bien des espoirs .

Or la sismologie solaire (comme celle de la Terre) est étroitement liée à nos connaissances sur la structure interne du Soleil , au moins sur celle des régions (jusqu'à plusieurs centaines de milliers de kilomètres de la surface , un tiers du rayon environ) qui sont dominées par les phénomènes convectifs . Si bien que toute amélioration de nos connaissances sur la sismologie solaire est une amélioration de notre connaissance des régions internes du Soleil (jusqu'à une distance du centre égale à 2/3 du rayon) . Toute découverte concernant la sismologie stellaire , rendue possible par les moyens spatiaux , sera , n'en doutons pas , un auxiliaire précieux pour les théoriciens des intérieurs stellaires , donc aussi de l'évolution des étoiles .

D'ores et déjà , des études ont été faites depuis le sol . Ainsi 53 Per (une étoile de type O4,5 V) a-t-elle manifesté des oscillations non radiales , variables

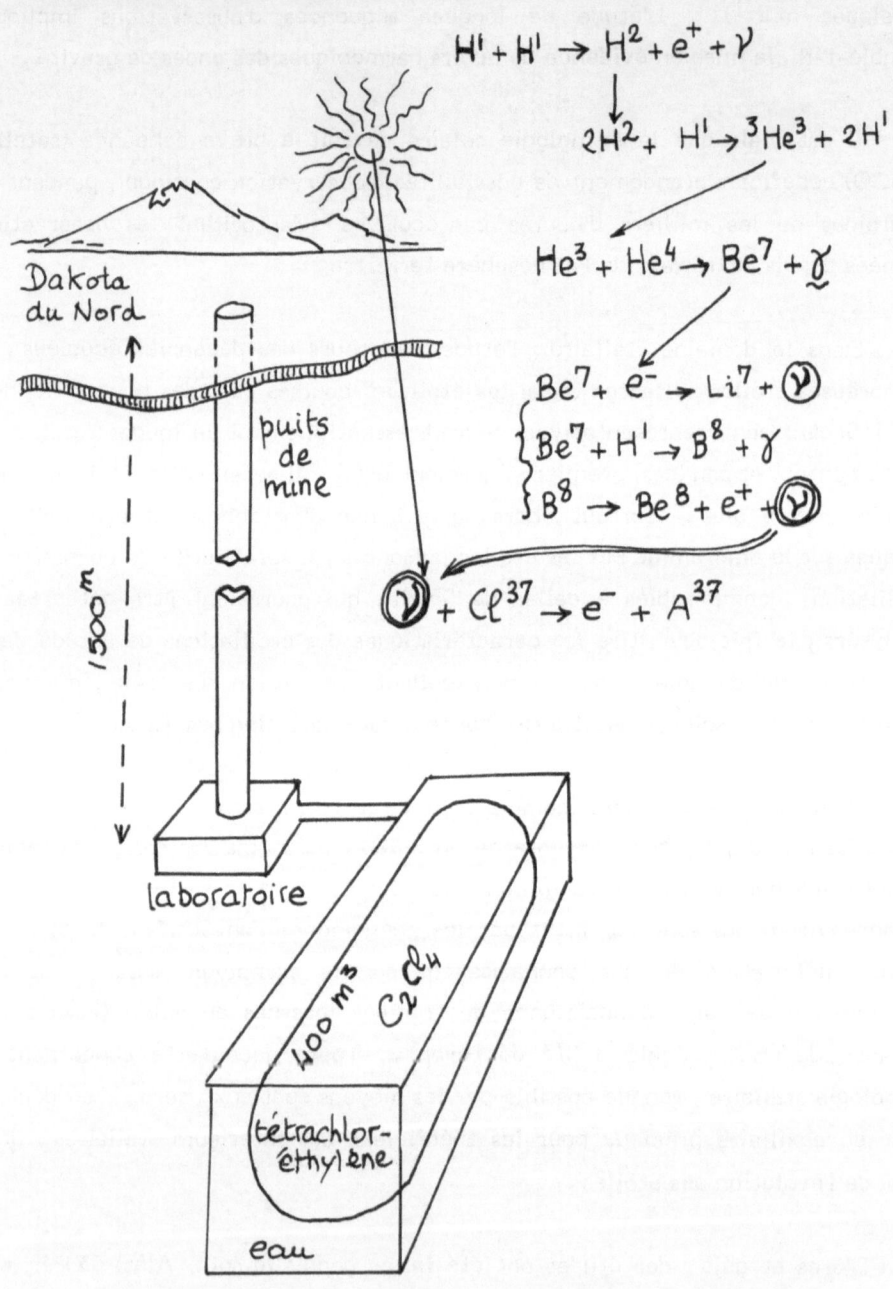

$$H^1 + H^1 \rightarrow H^2 + e^+ + \gamma$$

$$2H^2 + H^1 \rightarrow {}^3He^3 + 2H^1$$

$$He^3 + He^4 \rightarrow Be^7 + \gamma$$

$$Be^7 + e^- \rightarrow Li^7 + \textcircled{γ}$$
$$Be^7 + H^1 \rightarrow B^8 + \gamma$$
$$B^8 \rightarrow Be^8 + e^+ + \textcircled{γ}$$

$$\textcircled{γ} + Cl^{37} \rightarrow e^- + A^{37}$$

Dakota
du Nord

puits
de
mine

1500 m

laboratoire

400 m³ C_2Cl_4

tétrachlor-
éthylène

eau

d'ailleurs d'une époque à l'autre ; d'autres étoiles (22 Ori , ι Her , 10 Lac) , de type OB , ont un comportement analogue .

Il s'agit donc d'un domaine a priori extrêmement fructueux , qui doit bénéficier de la recherche spatiale (sur des engins constamment éclairés par le Soleil) , et qui doit avoir d'énormes conséquences sur la connaissance de l'intérieur des étoiles .

7 . La Neutrino-Astronomie

Il en va différemment de la neutrino-astronomie . On sait que certaines des réactions de fusion des protons au centre du Soleil produisent des neutrinos d'une assez grande énergie . Ces neutrinos , produits dans les régions centrales du Soleil , ne traversent ensuite que des régions de l'espace (intérieur solaire , milieu interplanétaire , Terre , planètes) pratiquement transparentes pour eux . Seuls un petit nombre d'isotopes capturent ces neutrinos . Ainsi en est-il pour le Cl^{37} , isotope qui constitue 25% du chlore chimiquement extrait de l'eau de mer . Cet élément se trouve dans de nombreux composés chimiques bon marché , comme le tétrachloréthylène , solvant industriel . Ainsi Davis , depuis plus d'une décennie , a-t-il étudié le comportement d'une cuve de 400 000 litres de ce solvant , immergée elle-même dans de l'eau (pour la protéger des neutrons produits par la radioactivité des roches) et localisée dans une mine d'or du Dakota du Sud , à

Figure 12   Solar neutrino-astronomy

Schematical description of the experiment conducted by Davis and his colleagues ; at the bottom of an old abandoned gold mine in South Dakota , the commercial solvant C_2Cl_4 is stored in a 400 m^3 tank , surrounded by water . The water acts as a screen against the neutrons emitted by the radioactive rocks ; the depth of the mine (1 500 m) protects the detector against cosmic rays . A current of helium takes away the gaseous argon A^{37} formed in the tank , and allows one to measure the number of captured neutrinos. The neutrino-productive reactions occuring in the Sun are indicated as well as the capture of the neutrinos by Cl^{37} .

1 500 m sous terre (ce qui la préserve des rayonnements cosmiques , d'énergie faible par rapport à celle des neutrinos) . Le Cl^{37} capte les neutrinos , et les transforme en A^{37} gazeux , qu'un courant d'hélium balaie , et permet de mesurer aussitôt . La mesure (qui tient compte évidemment de la faible durée de vie de l'A^{37}) montre que le nombre de neutrinos mesurés est d'au moins **trois fois inférieur** à celui que permettent de calculer les modèles classiques de l'intérieur du Soleil .

L'interprétation de cette différence flagrante résulte bien certainement du caractère inadéquat de ces modèles , ou de la théorie des neutrinos . L'existence d'oscillations de neutrinos massifs , réduisant le nombre des neutrinos életroniques à 1/3 du nombre total de neutrinos produits , n'est pas une hypothèse encore très bien confirmée par les expériences de laboratoire ; au surplus , elle semble quantitativement insuffisante . L'évolution du Soleil , sur quelques 10^6 ou 10^7 ans (temps mis par l'énergie lumineuse à se propager , par absorptions et émissions successives , de l'intérieur du Soleil à l'extérieur - à comparer avec les 8 minutes que mettent environ les neutrinos) parait peu vraisemblable , et cette explication semble , de toute façon , aller dans le mauvais sens . Plus cohérente est sans doute l'hypothèse de Schatzman et Maeder , qui , en suggérant l'existence de phénomènes de diffusion turbulente , ramenant vers le centre de l'étoile , où l'hydrogène se "consume" , de l'hydrogène frais , permettent , en diminuant la proportion de l'élément He^4 , de diminuer le taux de production des neutrinos ; ce processus a l'advantage d'expliquer aussi , de façon quantitative , le rapport He^3/He^4 observé comme anormalement élevé dans les rayons cosmiques d'origine solaire .

Il est évident qu'un engin spatial ne pourrait , en aucun cas , être aussi efficace que le dispositif de Davis , pourtant déjà insuffisant ; il est non moins évident que pendant des décades , la neutrino-astronomie restera confinée au Soleil: voilà un exemple où le développement de la technique au sol (et même en sous-sol) permettra sans doute (grâce à des cuves contenant du gallium) de mieux comprendre la physique des régions internes du Soleil: mais ce sera aux théoriciens seuls de nous dire si cette compréhension peut , et à quelles conditions , s'étendre aux processus à l'oeuvre dans les régions centrales des étoiles .

8 . Conclusions

Nous avons examiné seulement un petit nombre de problèmes ; de nombreux autres ont été laissés de côté: les problèmes posés par le caractère binaire de certaines étoiles , l'homogénéité de la surface stellaire (est-elle "tachée"?) , comment fonctionne la rotation différentielle , les mesures du magnétisme stellaire , etc .

Mais ces quelques exemples nous suffisent à conclure .

Le Soleil a certes dans nos préoccupations une position privilégiée . Il est tout d'abord **notre étoile** , celle où nous vivons ; et qui refuserait au Soleil ses frontières naturelles , celles du système solaire lui-même ?   De ce fait , les interactions entre le Soleil et la Terre jouent sur la physique de notre planète un rôle essentiel (voir ci-après l'exposé de Roger Bonnet) . De plus , il est pour les terriens un magnifique domaine d'études ; il peuvent y étudier des plasmas chauds , des gaz optiquement épais , et des réactions thermonucléaires (que nous préférons tous ne pas voir se dérouler de façon trop étendue sur notre planète) ; il peuvent utiliser son champ de gravitation pour vérifier les prévisions des théories relativistes ; pour nous , c'est donc aussi un **laboratoire** irremplaçable .

C'est aussi , on l'a dit sur tous les tons , une étoile banale , que l'on peut étudier comme telle , et qui nous a fait découvrir beaucoup de choses sur la physique stellaire .

Mais nous sommes en train de découvrir quelque chose d'autre: c'est que , si le Soleil est une étoile , les étoiles sont des soleils: leur étude , avec les moyens puissants dont nous disposons , notamment grâce à l'instrumentation embarquée sur des engins spatiaux importants , permet en quelque sorte d'expérimenter , de faire varier les paramètres , de voir ce qui arriverait au Soleil si l'on modifiait son rayon , ou sa masse , ou sa vitesse de rotation , ou son champ magnétique . Ayant commencé ce demi-siècle par l'étude du Soleil , comme substitut (au point de vue de l'étude de phénomènes importants , les phénomènes "actifs" notamment) de l'étude des étoiles , nous le terminerons en développant les études stellaires , comme le complément nécessaire à la compréhension physique des observations du Soleil. Peu à peu , physiciens solaires et stellaires sont en train de sortir de cette époque de douloureuse ségrégation , imposée par l'application de méthodes différentes à des objets regardés différemment . Ils nous donneront un outil solide , qui permettra d'aborder avec plus de sûreté les problèmes de l'évolution galactique , voire de l'évolution universelle , problèmes dont la solution est aujourd'hui encore très fragmentaire et très primitive .

9 . Bibliography

It is quite difficult to establish a reasonably well-balanced bibliography for such a vast group of subjects . I think it is better to refer the reader to recent books or symposium proceedings covering the relevant topics , and to contributions of the author , which , some of them in english , more or less form the basis of the present paper .

Proceedings of recent meetings

1.   Deuxième Assemblée Européenne de Physique Solaire , Toulouse , Mars 1978 , "Pleins Feux sur la Physique Solaire" , J. Rösch and S. Dumont (eds) , 1978 , CNRS , Paris .

2.   Proceedings of the S.O.T. Symposium , 24-25 Jan. 1980 , NASA , Goddard Space Flight Center .

3.   Fifth European Regional Meeting in Astronomy , Liège , July 1980 , "Variability in Stars and Galaxies" , Inst. Astrophys. , Liège , Belgique .

4.   Nato Adv. Stud. Inst. Symp. , Sept. 1980 , "Solar Phenomena in Stars and Stellar Systems" , R.M. Bonnet and A.K. Dupree (eds) , 1981 , Reidel , Dordrecht , The Netherlands .

5.   Proceedings of the Japan-France Seminar on Solar Physics , Oct. 1980 , "The Active Sun as a Star" , F. Moriyama and J.C. Hénoux (eds) , 1982 , Tokyo .

6.   Third European Solar Meeting of the E.P.S. , Oxford , Avril 1981 , "Solar Activity" , S. Jordan (ed) .

7.   Sixth European Regional Meeting in Astronomy , Dubrovnik , Oct. 1981 , "Sun and Planetary System" , W. Fricke and G. Teleki (eds) , Reidel , Dordrecht , The Netherlands .

8.   "The Sun as a Star" , 1981 , S. Jordan (ed) , NASA-CNRS Monographs series on Nonthermal Phenomena in Stellar Atmospheres , CNRS , Paris , NASA , Washington .

Author's recent papers dealing with the topics covered

1.   Pecker , J.C. , see ref. above 5 ("The Active Sun as a Star") , ref. 6 (concluding remarks) , ref. 7 (invited review paper "The Star Sun") , ref. 4 (preface , together with the other advisors and coordinators of the Series) .

2.   Pecker , J.C. , Thomas , R.N. , 1976 , Solar Astrophysics , "Ghettosis from , or Symbiosis with , Stellar and Galactic Astrophysics?" , Space Sc. Rev. , 19 , 217-243 .

3.   Pecker , J.C. , 1979-1980 , "Le Vent Solaire" , in Planetarium , Planetarium ed. , Bruxelles , p. 81-93 .

4.   Pecker , J.C. , 1980 , Forty Years of Research , "The Sun as a Whole ; its Photosphere" (unpublished invited discourse at the occasion of the 40th anniversary of High Altitude Observatory , Boulder , Colorado) .

5.   Pecker J.C. , 1980 , "Le Principe et le Développement par Bernard Lyot du Coronographe" , C.R.Acad.Sc. Paris , T.291 , Vie Académique , 17 Nov. 1980 , p. 79 .

6.   Pecker , J.C. , 1982 , "L'Atmosphère Solaire , du Modèle à la Physique" , Essais et Conférences , Collège de France , P.U.F. , Paris.

CHAPTER II

SOLAR-TERRESTRIAL INFLUENCE

by R.M. Bonnet

Laboratoire de Physique Stellaire & Planétaire

C.N.R.S.

91370 Verrières le Buisson , France

Summary

The sun extends far away : the earth is in the sun .

It is therefore no surprise that the sun has a direct influence on the earth . The sun's influence on the earth and the planet exerts itself through gravitation , particles from the solar wind , the magnetic field and radiation . Nearly all the related effects have been studied or discovered mainly because of advances in space research . 25 years of space research have indeed revealed the existence of the solar wind and the structure of the interplanetary medium between the earth and the sun as well as the structure of the solar magnetic field , and the detailed shape and evolution of the earth magnetosphere .

The interaction between the earth's atmosphere and solar radiation is strongly dependent upon the portion of the spectrum that is considered . The upper layers are directly influenced by the ultraviolet whose intensity is strongly dependent upon the degree of activity of the sun .

R. M. West (ed.), Understanding the Universe, 37–71.

How far down does this influence extend in proximity to the ground and to what extent is the climate of the earth influenced by the variation in the total radiative output of the sun are questions of major interest to humanity . Not only do they relate to our daily activity (the weather , the climate) but also to the evolution of life .

These various problems are discussed . Historical aspects as well as the most recent results provided by space research are described .

1. Introduction

The study of the relationships between our sun and our earth is not what we would call a new science . The Ancients had already identified in the sun the prime source of energy of life . The celebration of the winter solstice at Christmas time marks the happiness of seeing the course of the sun in the heavens reassured once more . Curiously enough , we may find today in the behaviour of many people the remnants of such ancient and naive beliefs such as the occurrence of solar eclipses that was attributed to the unfathomable will of some obscure god . How many of us have not been approached by friends , or by people in the street at times of particularly cold winters or summers or of unusual heat waves , who were concerned about a disorder of the solar machinery or by an extraordinary level of activity of the sun ? Many of us may smile to these naive remarks or interrogations . However , do we really know the answer ? Do we have in our hands the means of evaluating both in a qualitative and in a quantitative way the influence that the sun may exert on our daily life or on its evolution ? In the following we will attempt to give an answer to this question .

Historically several phenomena encountered on the earth have been thought to have their origin in the sun . For example , as early as 1860 the British physicist R.C. Carrington had noticed the coincidence between intense **eruptions** on the sun and the occurrence , a few hours later , of **polar aurorae** . The correlation between the geomagnetic perturbations and the number of sunspots , an index very commonly used to measure the degree of solar activity , was found also to be very high . We owe to the two Scandinavians , Stormer and Birkeland , the first suggestion that aurorae and geomagnetic perturbations are caused by electric particles ejected by the sun .

Soon after 1900 it became progressively clear that the propagation of radio waves was influenced by a layer of electric particles in the upper atmosphere (the **ionosphere**) . But it was only around 1925 that the origin of these particles was suggested to be the ionization of the species of the atmosphere by ultraviolet radiation from the sun . Unfortunately , the ultraviolet solar spectrum as observed from the ground was devoid of any energy for the very reason that this energy was absorbed in the upper atmosphere of the earth .

These facts allow us to stress the role played by space research in the efforts of man to properly evaluate the effect of solar phenomena on our environment .

This role cannot be underestimated . The first space experiment was done by a group at the Naval Research Laboratory in the USA on October 10 , 1946 . It consisted of launching an ultraviolet spectrometer above the earth's atmosphere by means of a V 2 rocket . The first ultraviolet spectrum of the sun , below the cut-off due to the absorption by the atmosphere , was obtained on that historical day. It showed the importance of the absorption by ozone and measured the vertical extent of this species in the earth's atmosphere (Baum et al , 1946). Since that time , many more rockets , satellites and balloons have been launched , among which a vast number were devoted to the study of the relationship between the sun and the earth .

The interest in this field has not vanished today , and nearly 36 years after the launch of the first rocket experiment , programmes are underway in several countries to try to elaborate more precise correlations , measure more precise parameters and get a clearer understanding of which are the crucial ones that may play a role in the delicate balance that exists between the temperature , the density , the chemical composition of our atmosphere and in the evolution of our living conditions : the weather and the climate .

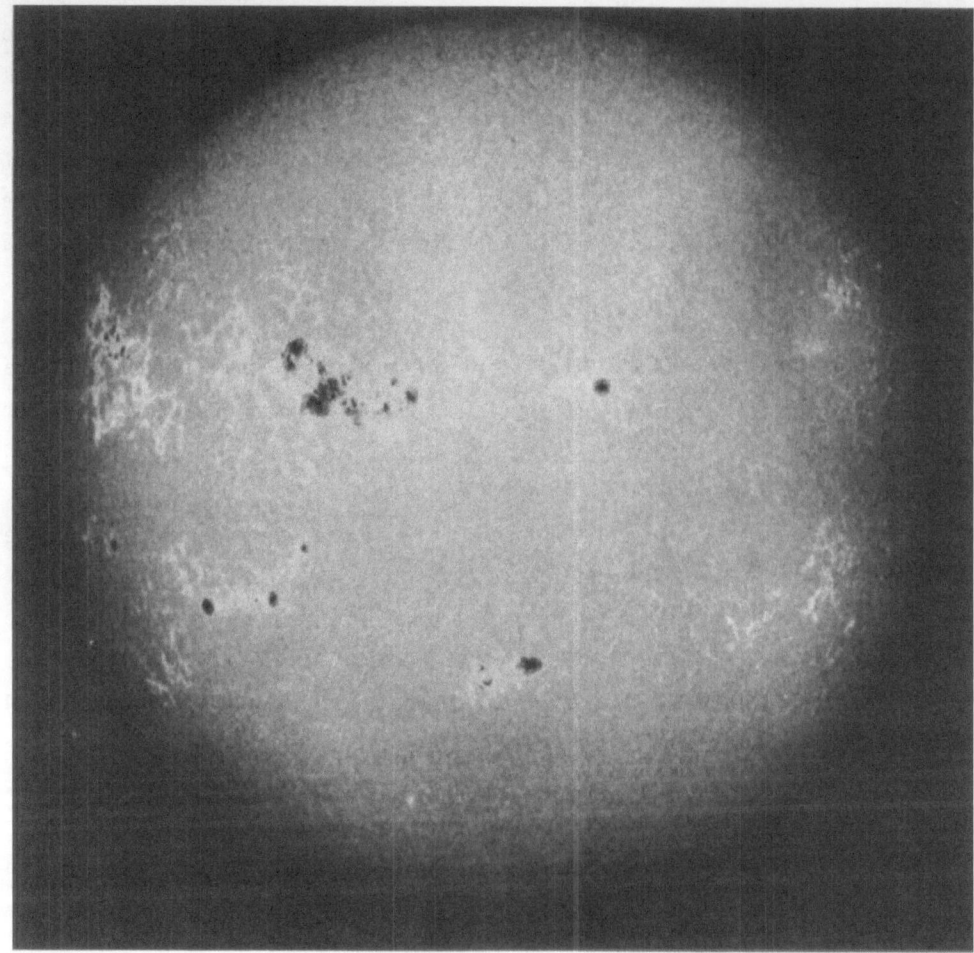

Figure 1   Broadband image of the solar disk obtained through an ultraviolet filter showing most of the characteristic features of the photosphere from which the light from the sun is emitted . The various sunspots (dark areas) and faculae (bright area) are easily distinguishable .   The photograph was obtained on 13 July 1982 , (photo : Laboratoire de Physique Stellaire et Planétaire) .

2. The Solar Influence

The vicinity of the sun to the earth manifests itself in three different ways :

- the earth is gravitationally coupled to the sun ;
- the earth is imbedded in the solar corona and is subject to the direct influence of its magnetic field and the bombardment of particles ejected from the sun , either continuously (solar wind) or at the time of flares ;
- the earth is constantly illuminated by the sun . It receives an energy equal to 1.367 kW/m^2 .

Of these three factors only the last two will be considered here . However , it will be shown that the characteristics of the earth's orbit around the sun that relate to the first factor may play a determining role in the behaviour of our past and near future climate.

Table I compares the amount of energy contained in the various solar factors that we identify as having the strongest influence on the earth . We immediately notice that radiation contains six orders of magnitude more energy than the particles and the magnetic field . We should therefore expect that this factor may exert the largest influence . In fact we will see that this is not so simple to ascertain .

TABLE I

COMPARISON OF ENERGY FLUXES AT ONE A.U. EXPRESSED IN ERGS/CM^2/S .

Radiation	1.36 million
Solar wind	2
Magnetic	0.01

Figure 1 is what we usually call an image of the solar disk , on which we easily distinguish the sunspots , and the facular regions . These features and their

Figure 2   This photograph is in reality a map of the small scale magnetic field of the sun at the surface of the photosphere . Black and white colours represent the two opposite polarities of the field . Notice the larger intensity and larger concentration of magnetic field in the regions corresponding to the spots and the faculae of Figure 1 . This map was in fact made at nearly the same time as the picture of Figure 1 (picture from Kitt Peak National Observatory) .

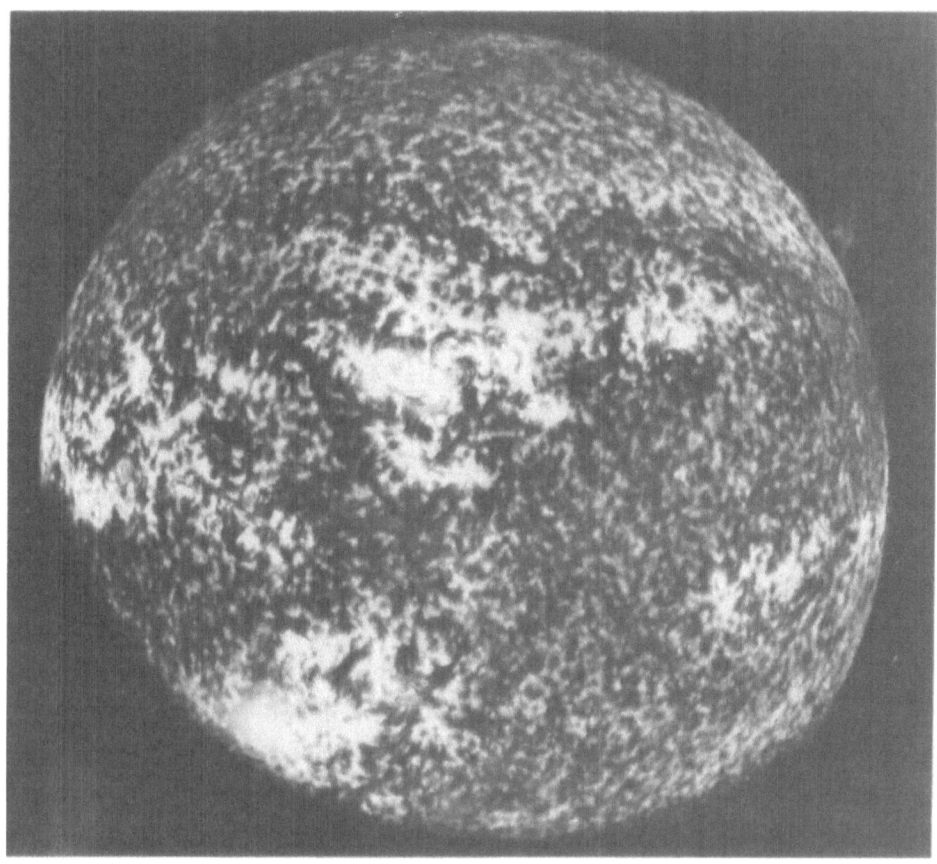

Figure 3   An image of the sun's chromosphere as obtained through a filter that lets only the Lyman alpha resonance light emitted by the hydrogen atom go through .   Most of the emission of light by the chromosphere is in the far ultraviolet which is absorbed in the upper atmosphere of the earth and is responsible for the existence of the ionosphere . Notice the rather inhomogeneous character of the emission which corresponds to regions of strong magnetic field (photo : Laboratoire de Physique Stellaire et Planétaire) .

distribution characterizes the emission of light from the so-called **photosphere** of the sun . We may expect that the variation of the number of sunspots and faculae and of their size may induce a correlative variation in the total radiative output of the sun . We will come to this point below in Section 5 .

Figure 2 is a magnetogram of the sun or in other words a map of the distribution and intensity of the solar **magnetic field** in the photosphere . It was obtained at the same moment as the picture of Figure 1 . We easily see that there is a close spatial coincidence between sunspots and faculae and the regions of intense magnetic fields . Most of the variability of the sun , at least on a time scale of a solar cycle , is due to its magnetic field . It imposes the distribution of active regions , it is at the origin of flares and is likely to be the cause of the existence of the solar corona . The magnetic field is created in the interior of the sun through a dynamo mechanism that results from the coupling between the movements of convection and of rotation . It is amazing to think that several phenomena on earth which are caused by solar variability are in fact related to phenomena that occur beneath the photosphere in the convection zone of our star .

Above the photosphere are the chromosphere and the corona . These are regions where the temperature increases outward and the density decreases rapidly with altitude . As a result , the gazeous pressure decreases more rapidly than the pressure exerted by the magnetic field , and the field in fact models more and more easily the structure of the chromosphere and even more that of the corona. The **chromosphere** in particular (Figure 3) is responsible for most of the emission of UV radiation that in itself is responsible for the existence of the ionosphere and is absorbed in the earth's atmosphere . The distribution of UV radiation on the color disk is so inhomogeneous that we may expect a strong variability of the solar output in this region of the solar spectrum .

Above the chromosphere is the **corona** easily seen at the time of eclipses but thanks to X-rays and radio observations we now are able to observe it in greater detail , not only above the visible limb but also on the disk of the sun .

We owe to the American space station **Skylab** the most extensive studies of the solar corona both in X-rays and in white light . In particular , the X-ray pictures have revealed the extreme complexity of the magnetic structure of the corona and the presence of spatially extended intensity depressions (called **coronal holes** for these very reasons) . It has been shown also that coronal holes correspond to regions where the magnetic field lines are open (contrary to their general loop shapes) closing far away from the sun in interplanetary space . Such open field lines are quasi permanently observed at the solar surface . It has also been shown that the solar wind blows preferentially out of these regions .

With its white light coronograph , Skylab was able to observe the disruption of magnetic structures and to follow the ejection of vast amounts of matter (electrified gases) and particles (electrons and ions) into interplanetary space . These particles after their journey reach the earth a few hours or days later , depending on their velocity , and are responsible there for several spectacular phenomena .

The small scale solar magnetic fields influence the chromosphere and the corona . The solar wind , as we have seen , is influenced by the large scale structure field . This structure can be deduced by averaging the small scales (Figure 2) and by computing (Svalgaard and Wilcox , 1978) the radial component of the field . The result is a figure that is made of four sectors of alternately positive and negative polarities . The spatial extension into interplanetary space is shown on Figure 4 from Wilcox (1980) . The warped surface thereon represents the separation between the positive and the negative polarities. The helical shape of this surface is due to the solar rotation and to the expansion of the corona. The field is in fact carried out into space by the plasma ejected by the solar wind and the combination of the two motions (rotation and expansion) results in this funny shape . The earth is swept by sectors of alternative opposite polarities during solar rotation . The discontinuous jump in the field at the crossing of the border between two sectors may , as we can guess , induce electric perturbations the same way an alternator would in an electric generator . Near sunspot minimum

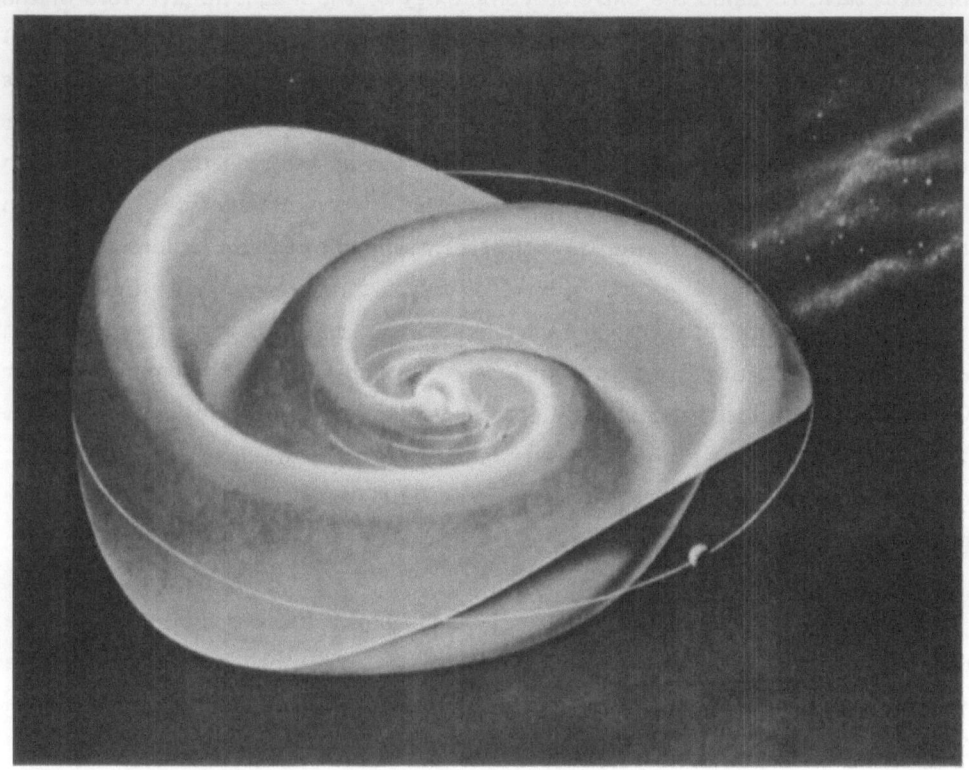

Figure 4   An artist's impression of the warped surface which separates in interplanetary space the two regions where the magnetic field has opposite polarities.   Notice the helical shape of the surface that results from the combined motions of rotation of the sun and expansion of the magnetic field . The orbit of the earth is also seen in the picture . We realise that the earth is swept alternatively by sectors of opposite polarities (figure from Wilcox , 1980) .

the warped sheet is fairly close to the equatorial plane of the sun with the extent in latitude being 15 to 20 degrees north and south (as in the Figure). Near maximum , the sheet can reach latitudes of ≃ 50 degrees north and south .

We shall now analyze in more detail the consequences due to the fact that the earth is embedded in the solar corona .

3. The Consequences for the Earth of Being Embedded in the Outer Solar Corona

At the distance of one astronomical unit the physical characteristics of the solar corona differ markedly from what they are in the close vicinity of the sun . The average density of particles per cubic centimeter is 10 (mostly protons and electrons) . In comparison there are some of 27 billions of billions of molecules per cm^3 in the earth atmosphere at sea level . By laboratory standards the solar corona is what we would call an ultra high vacuum . The magnetic field (0.00007 gauss) is only 1/10,000 of the strength of the earth magnetic field .

How can such a tenuous medium have any influence on the terrestrial environment and on the atmosphere of the earth ? It is in fact responsible for spectacular manifestations and this is essentially because of the effect of the earth's own magnetic field .

We owe to the American satellite **Explorer 1** the discovery in 1958 of the presence of concentrations of energetic particles around the earth , the so-called **van Allen belts** (so-named from the scientist who made the first observations) , and we owe to the Soviet probe **Luna 3** the first direct observations of the solar wind and the first measurements of its characteristics . Table II summarizes the average characteristics of the expanding corona in the immediate vicinity of the earth .

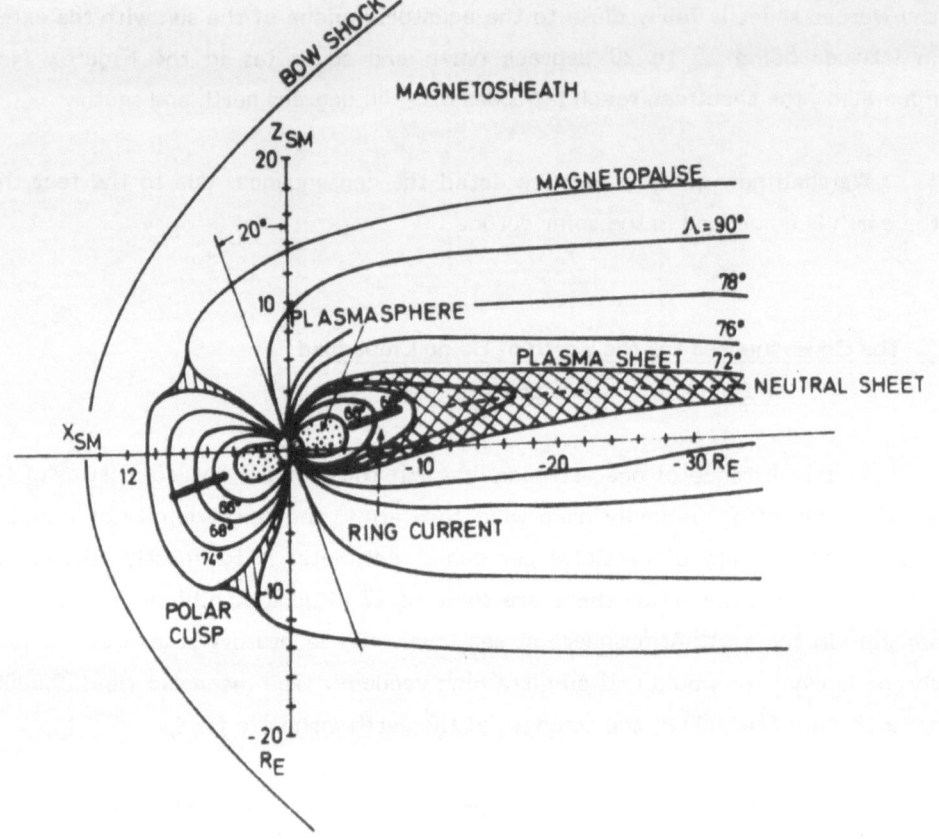

Figure 5   The magnetosphere of the earth is defined by the interaction between the solar magnetic field and the terrestrial magnetic field . At the interface between the two , a bow shock exists which separates the regions where the solar field and the earth field are dominant .

Highly energetic particles can penetrate through the shield of the magnetosphere and eventually be trapped in the van Allen belts (plasmasphere) . They can also precipitate near the poles down through the polar cusps which offer a natural way of entry .

Particles from the solar wind are stored in the tail of the magnetosphere (plasma sheet) from where they precipitate into the polar cusps at times of magnetospheric substorms and create polar auroraes .

TABLE II

CHARACTERISTICS OF THE SOLAR WIND AT THE EARTH ORBIT

Composition	Protons , electrons , helium , ions 5-10%
Temperature	10 000 K , 100 000 K
Velocity	450 km/s (average)
Density	10 particles/cm^3
Magnetic field	about 1/10,000 of the earth magnetic field

3.1 The Earth's Magnetosphere

The first interaction between the sun and the earth occurs at about 14 earth radii , the distance where the kinetic pressure of the solar wind starts to be counterbalanced by the pressure , exerted by the earth's magnetic field . The effect is the creation of a **bow shock** (Figure 5) which defines the border of the magnetosphere , a water-drop shaped cavity that has remarkable properties , so remarkable in fact that they have created a totally new area of research and are still today the rationale for a large number of space programmes .

The global effect of the magnetosphere is to deviate the particles of the solar wind from their unperturbed trajectories , protecting the earth and the living creatures thereon from a fatal bombardment . The earth's magnetic field exerts a force on the electrons and the protons , and constrains them to either gyrate around the lines that delineate the field or to drift , depending upon their direction and their velocity .

i)   The case of the high energy particles

Highly energetic particles originating from solar flares can penetrate down to 2 to 5 earth radii . This is easier in the vicinity of the earth's poles .

They are trapped in the so-called **radiation belts** , where they can oscillate back and forth for several months and even years . The trapping mechanism involves genuine nuclear reactions in the earth's atmosphere , during which neutrons are created which later disintegrate in a proton and one electron . These can either leave the magnetosphere if their trajectory corresponds to a weak field line , or be trapped in the other case . Those which have neither the right velocity nor the right direction can reach altitudes as low as 30 to 20 km and then disappear , depositing their energy in the form of heat .

ii)   The case of solar wind particles

The less energetic particles from the solar wind are driven into the tail of the magnetosphere where they are stored , and where their energy accumulates . This process is suddenly interrupted , during what is called a **magnetospheric substorm** , through a series of complex mechanisms the net result of which is to liberate the particles by injecting them into the upper atmosphere , preferentially near the poles of the earth on the night side , where their energy is dissipated into heat . The particles which are accelerated by the magnetospheric substorm in the vicinity of the magnetic poles collide with the atoms of the atmosphere and it is the fluorescence induced by these collisions that create spectacular luminous manifestations : the **aurorae borealis** .

In summary , the earth's magnetosphere has two principal properties :

- it prevents the solar wind particles and in particular the more energetic ones emitted by the sun at the time of flares from reaching the earth's surface ;

- it stores the energy carried by these particles and suddenly , during magnetospheric substorms , releases it in the auroral zone and at the level of the ionosphere ;

3.2 The Net Effect of the Solar Particles on Our Immediate Environment

Because of the focussing effect of the polar cusps most of the phenomena linked to the particles of solar origin occur in the auroral zones from where they are transferred to other areas .

i)   Heating

As we have just seen solar energetic particles can penetrate directly into the atmosphere while solar wind particles penetrate through the tail of the magnetosphere and are injected into the polar cusps . In both cases their energy is dissipated into luminous phenomena and heat . The energy that is involved in these phenomena is only 1/10 000 the total energy carried by solar radiation .

ii)   The effect on the ozone layer

Another effect of solar particles , probably of a more consequent importance , is that which affects the ozone layer . Ozone which results from the association of one molecule O_2 and one atom of oxygen O provides an efficient screen against the very harmful solar ultraviolet radiation that dissociates it into its two basic constituents according to the reactions :

$$O_2 + O \rightarrow O_3 \text{ (ozone)}$$

$$O_3 + \text{ultraviolet light} \rightarrow O + O_2$$

The chemistry of ozone is quite complex but is essentially controlled among other factors by nitric oxide NO . When they penetrate to the level of the mesosphere (80 to 50 km) the solar protons easily dissociate molecular nitrogen and oxygen that , in their atomic form , associate to give nitric oxides whose net effect is to destroy O_3 according to the process shown on Figure 6 . During at least two solar proton events evidence has been observed of strong ozone depletions at high latitudes . Although the total ozone content in the atmosphere is not modified , the local depletion may last for several days or even weeks before the atmosphere comes back to its original state (Mitra , 1980 , Keating , 1981) .

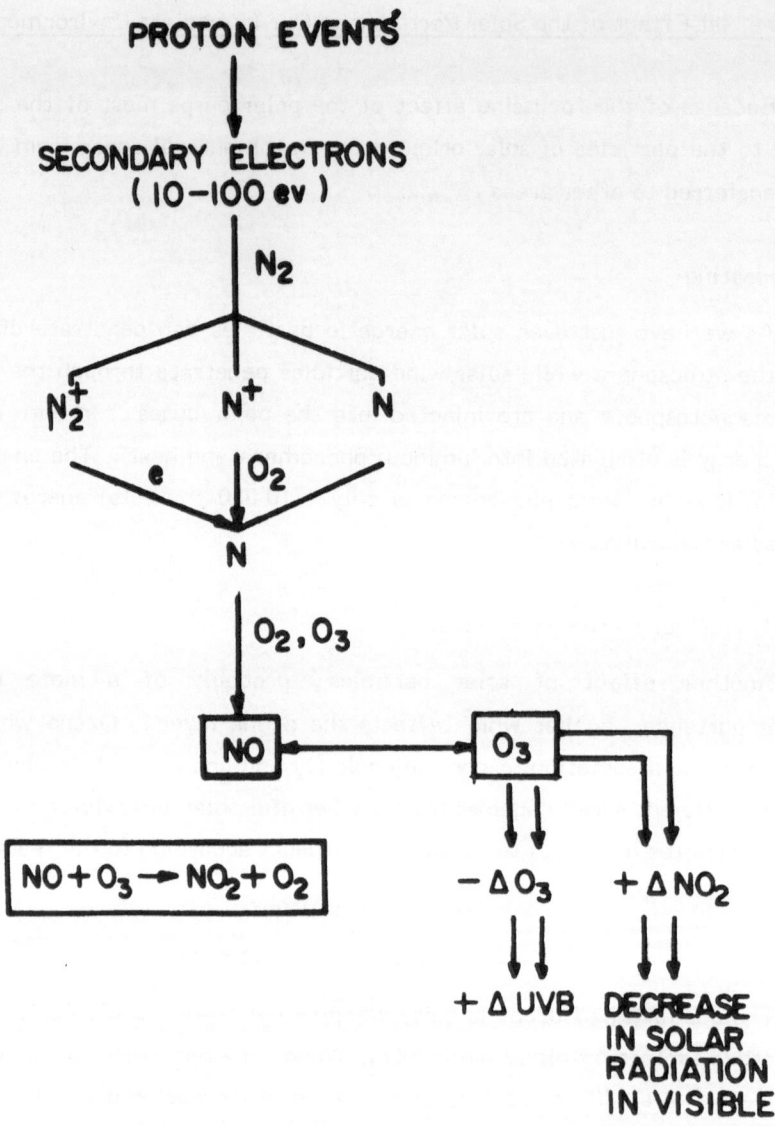

Figure 6   Main sequence of reactions, involving nitrogen and oxygen compounds, triggered in the upper atmosphere by the arrival of high energy protons from solar flares. The creation of NO results in a depletion of the ozone content at the location of penetration of the protons. Ozone depletion results in excess ultraviolet radiation transmitted by the atmosphere (figure from Mitra, 1980).

iii)   Electric phenomena

When they penetrate into the atmosphere , magnetospheric particles create electric currents at the level of the ionosphere ($\simeq$ 100 km). This is the so-called **auroral electrojet** which under the force induced by the earth magnetic field , drifts and modifies the atmospheric circulation . Highly energetic particles that penetrate down to 20 or 30 km also ionize the atmosphere . Through vertical circulation the charged particles may reach the troposphere and modify the electric conductivity of the clouds . This may be the source of thunderstorms and heavy rains although the detailed mechanisms involved in this process are far from being understood yet .

iv)   Electric black-out

The most spectacular effects however occur near the surface of the earth in the vicinity of the auroral zones .

There , the electric currents produced by the ionization of the atmosphere and which are highly variable and intense , induce electric fields beneath the earth's surface . These fields that can reach an intensity of 5 volts/km , and produce dramatic effects on any conductor of large dimension such as pipe lines or telephone wires or high voltage lines . This is probably the origin of the famous New York black-outs that occurred in May 1969 , August 1972 , and more recently in April 1981 and which were coincident with very energetic solar flares . The detailed mechanisms involved by these phenomena are now well understood and protective measures can be taken in order to avoid their repetition and their disastrous consequences .

3.3 Geomagnetic Perturbations

Although much less energetic , the modification of atmospheric electric currents as a consequence of the variations of the interplanetary magnetic field (Wilcox and Scherrer 1981) has been recently studied and it is suggested that it may have a meteorological influence . The effect would be particularly important in regions of intense tropospheric circulation . The tone of caution of these sentences implies that we are not yet in a position to prove unambiguously that we have indeed a direct relationship between the crossing of a magnetic sector and the meteorological phenomena mentioned here .

4. The Interaction Between Solar Radiation and the Atmosphere of our Planet

Of the three energy sources of Table I, i.e. particles , fields , and radiation , the latter is quantitatively the most important . We should consequently expect that the radiation from the sun , and its variation , is a source of direct perturbations of our environment .

The basic interaction of radiation with the atmosphere is through the absorption of photons $(h\nu)$ by the constituants C of the atmosphere :

$$h\nu + C \rightarrow C^* .$$

Here , C^* represents any state of a particle of constituant C after disappearance of the photon by either dissociation , excitation , or ionization . In this process the energy carried by radiation is either transformed into heat or re-radiated at the same or any other wavelength .

Any consideration of the effect of solar radiation on the earth atmosphere must take into account three quantities :

- the intensity of solar radiation
- the ability of atmospheric species to absorb this radiation ;
- the quantity (abundance) of atmospheric species .

The first and the third quantities are not independent . Therefore we should not in principle expect any simple relation between the absorbed energy rate and the variation of solar radiation . This is one difficulty in the search for evidence of atmospheric phenomena directly related to solar variability .

Another difficulty comes from the all too obvious scarcity of long term measurements : this makes experimental comparisons rather difficult . This is why nearly all long term trends of atmospheric parameters and their relationships with solar radiation fluctuations are based half on measurements and half on theory .

Different portions of the solar spectrum affect the atmosphere at different altitudes. Table III shows the correspondence between the various spectral regions and the ranges of altitude in the atmosphere, in which the corresponding radiative energy is absorbed. While the far ultraviolet and the X-rays mostly react with species in the upper atmosphere, only the "visible" light (which however contains 99% of the energy) reaches the ground unabsorbed. Under such circumstances one may wonder if the 1% left has any influence at all. The answer may be found in Table III which compares the kinetic energy of the atmosphere above a given level with the solar energy which is absorbed above that level. It can be seen that the two are not too different, particularly in the upper layers.

TABLE III

ATMOSPHERIC COLUMN ENERGY CONTENT
(24 hour illumination)

Altitude (km)	Absorbed energy at this Altitude (ergs/cm^2)		Kinetic energy (ergs/cm^2)		Wavelength range (nm)
100	0.43	million	0.25	million	5 - 105
85	1.7	million	2.5	million	5 - 175
50	7	million	700	million	5 - 200
15	1.4	billion	70	billion	5 - 300
0	118	billion			5 - 1000

Broadly speaking, the degree of variability of the solar spectral irradiance due to activity increases inversely with the wavelength. In other words, the further in the ultraviolet is the radiation, the more variable its intensity. Following Table III, we may therefore expect that the higher layers of the atmosphere will reflect more clearly the fluctuations of solar radiation due to solar activity than the lower ones. In effect, solar UV radiation strongly influences the structure of the ionosphere and of the thermosphere.

4.1 The Ionosphere

The far ultraviolet and in particular the very intense Lyman alpha line (the line emitted by hydrogen atoms at their resonance frequency) is responsible for the existence of the ionosphere through the dissociation and ionization of molecular nitrogen and oxygen .

To the question what is the net effect of solar ultraviolet variability on the atmosphere above 100 km the answer is not completely straightforward since it affects at the same time both the density and the temperature , as we show in the two following examples .

However , ultraviolet radiation can be considered to be the most important heat source of the upper atmosphere since it influences permanently 50% of the whole thermosphere . We have already indicated that particle precipitation and the dissipation of electric currents , in particular at high latitudes , are also a source of atmospheric heating .

i)   The effect on the density

It is well known that low altitude artificial satellites have limited life time due to the drag effect exerted on them by the atoms and the molecules of the upper atmosphere . In principle, if we know the density of atoms at the altitude of a spacecraft we are able to accurately predict the residual atmospheric drag effect and the precise moment when the satellite will enter the atmosphere . In reality since this drag effect is most important in the layers which are the most sensitive to variations of solar activity such predictions are far from being accurate as was obvious by the early burn out of the Skylab station above the Indian Ocean in 1979 and in the case of MAGSAT for which the predictions were short by nearly three months , because they were made by assuming a higher level of activity than was actually the case .

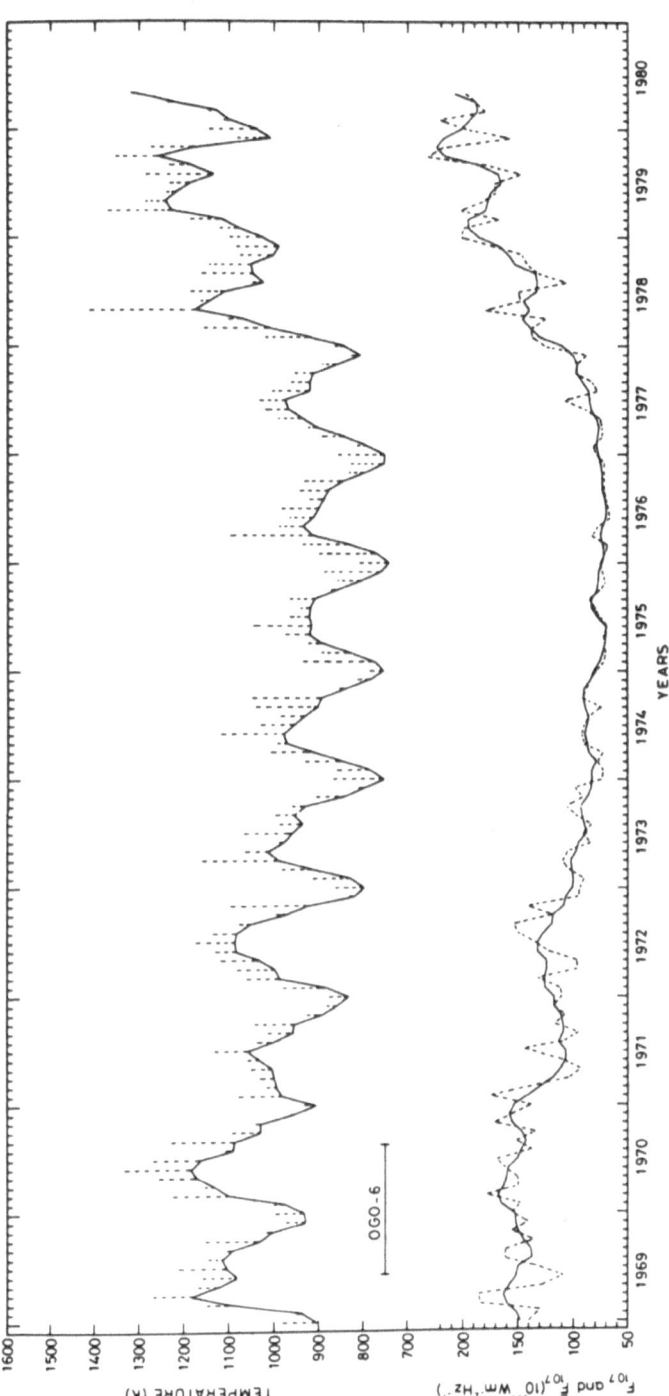

Figure 7 The variation of the temperature of the thermopause (upper curve) at 17 H (local time) for a region located at 52° north, from 1969 to 1980. Vertical dashed lines indicate temperature increases resulting from geomagnetic activity. Notice the remarkable modulation due to the seasonal variation of temperature and the general fitting with respect to the level of solar activity indicated on the lower curve by the variation of the solar radio flux at 10.7 cm. The horizontal line labelled OGO-6 indicates the time period over which the temperature was directly measured on-board the OGO-6 spacecraft. The values at other times are deduced from a semi empirical model (Kockarts, 1981).

ii)   The effect on the temperature

Figure 7 (Kockarts , 1981) , clearly shows the variation of the temperature
of the thermopause at mid European latitudes with the 11-year solar cycle .
The results of this figure are based on a semi-empirical model which bears
on the precise measurements of the OGO-6 satellite made in 1969-1970 .

These two examples stress the obvious necessity of knowing accurately the
value of the ultraviolet flux of the sun as a function of time if we want to
predict the behaviour of the upper atmosphere .   The corresponding
measurements can only be done by means of space borne instruments .

4.2 The Response of the Middle Atmosphere to Solar Radiation : the Ozone Layer

Because of the major importance of the ozone layer for our environment and
for our life we will mainly concentrate here on that portion of the atmosphere
extending from ≃ 20 to 60 km , i.e. the stratosphere and mesosphere , where the
concentration of ozone is at a maximum .

The thermal structure of the stratosphere and of the mesosphere is mostly
controlled by ozone . In addition to the destructive effect of NO (see Section 3)
inverse effects that increase the concentration of ozone result from the injection
of chlorofluor methanes , water vapour and carbone dioxyde (CO_2) in the
atmosphere by human industrial and military activity .

The possible effect of the solar activity cycle on the ozone concentration
has been a subject of controversy for more than half a century . No simple or
direct relation should be expected, given the great complexity of chemical
processes involved , the existence of large dynamical effects when particular
areas on the earth are considered and above all the lack of precise knowledge of
the degree of variability of the UV flux in the wavelength range of interest .
However , artificial satellites have made it possible today to perform **global
measurements** of the ozone concentration which , by smoothing out the effects of
local dynamics , allow a possible solar radiation effect to be detected .

Such measurements were made by the **Nimbus 4** satellite (Keating , 1981) . They are evidence of a temporal variation which follows with a time lag of about one month the variations of the solar flux at 10.7 cm . The increase in ultraviolet flux of 20% required to represent this variation is in close agreement with the measurements performed by the **Atmospheric Explorer** satellite . Both the time lag and the 3% variation observed in ozone concentration between solar maximum and solar minimum are in agreement with the theory .

There is also evidence of temperature effects with solar activity (Keating , 1981) .

How these effects compare with those of human origin resulting from the abuse of agricultural fertilizers and fluorochloromethane is a question yet to be answered . By all means it is necessary to know what the natural effect is , if we want to infer and hopefully correct the consequences of our agricultural and industrial activities .

4.3 The Response of the Lower Atmosphere to Solar Radiation

Of more immediate interest is the response of the lower atmosphere to solar fluctuations . Because of the perturbing effect of the albedo of the earth , the clouds , the oceans and of the atmospheric circulation , the quite good correlations observed at higher levels become more and more difficult to observe at lower altitudes . If we try to represent the variations of any parameter P of the atmosphere like density , or temperature by the following formula :

$$P = P_0 (1 - kR)$$

where R represents the index of activity of the sun (it can be either the sunspot number or the 10.7 cm solar flux) and P_0 is the unperturbed or minimum value of P , then k may vary by 3 to 4 orders of magnitude between 2 km and 400 km .

Only at altitudes larger than 5 to 8 km can any effect be reliably observed on the temperature of the atmosphere (Schwentek and Elling , 1981) . The response is more important in winter than in summer, probably because the atmosphere is less subject to vertical transport by convection or diffusion .

5. Have The Solar Variations Any Influence on our Climate ?

We already have a hint from the past two sections that it may not be so easy to establish the answer to this question . Taking into consideration the overwhelming importance of solar radiation in the overall energy balance of the earth (as shown on Table I) one would be tempted to say that our climate is totally influenced by the variability of the energy contained in that portion of the solar spectrum that reaches the earth's surface, i.e. the visible spectrum . On the other hand , we have just seen that the deeper we penetrate into the atmosphere the more difficult it is to find a visible correlation between solar fluctuations and the parameters characterizing the state of the atmosphere .

In the absence of any strong internal energy source (which is more or less the case for the earth) the temperature of a celestial body illuminated by the sun is obtained through a simple equation expressing that the energy absorbed by the planet is reradiated as from a body of temperature T . If S is the solar constant (or the total energy content of solar radiation) and R the radius of the earth this equation can be written as :

$$\pi R^2 S = \sigma T^4_{earth}$$

where σ is a constant derived from quantum mechanisms . With $S = 1.367$ kW/m^2 one finds that $T = + 5^\circ C$. This is cold compared to the actual, measured average temperature which is closer to $+ 15^\circ C$. The difference comes from the blanketing effect of the atmosphere which was neglected in our simple equation and which prevents most of the infrared radiation emitted by the earth from escaping into space . Atmospheric CO_2 and H_2O , carbon oxyde , water vapor and ammonia (NH_3) are the most efficient agents in this process . Conversely the icy polar caps and the clouds efficiently reflect a substantial portion of the solar energy into space . The balance between these two opposite effects may explain the difference when they are correctly taken into account .

The equilibrium formula above also allows us to determine what temperature variation may result from a variation of the solar constant S . We find easily $\Delta T_e/T_e = 0.25\ \Delta S/S$. In other terms a variation of 1% in S would result in a variation of $0.7^\circ C$ in T_e. However, this is again a rather crude evaluation because we have ignored most of the positive or negative feed-back effects such as the increased evaporation of the oceans which increase the blanketing effect, the larger albedo due to a larger cloud and ice cover etc. It is presently accepted that a variation of S of 1% would in fact result in an increase of T_e between $1^\circ C$ and $2^\circ C$ (Wigley, 1981). This is a relatively large effect and we easily realize the importance of knowing accurately the absolute value and the variations of the solar constant.

5.1 Evidence of Variation of the Solar Constant

The theory of the evolution of our star shows that in its early stages of evolution the total energy output of the sun was 25% lower than at present. The fact that no evidence of a subsequent low temperature of the earth has ever been detected constitutes the so-called **Faint Sun paradox**. The solution usually proposed is that of an increased blanketing due to the presence in the atmosphere during the early stages of the earth's evolution of a small amount of ammonia : only 10 to 100 ppm would be sufficient to solve the paradox.

Systematic measurements of S were undertaken only recently by Abbott in the USA between 1923 and 1954. Before that time, no measurements are available that would allow a survey of systematic trends. The low accuracy of Abbot's measurements did not allow to detect any variation with an amplitude less than 1%. This limit was mainly imposed by the perturbing effects of the earth's atmosphere. The advent of space research has offered a unique way of measuring the solar constant with an accuracy which has reached the incredible value of a fraction of a percent. The best value obtained as of today for S is 1.367 ± 0.002 kW/m^2. (Figure 8 ; Fröhlich and Brusa, 1981).

The most recent measurements performed on both the two satellites **Nimbus 7** and **SMM** reveal relatively large diminuations of S of $\approx 0.1\%$ that may last several days. These dips have been proved to be directly correlated with the

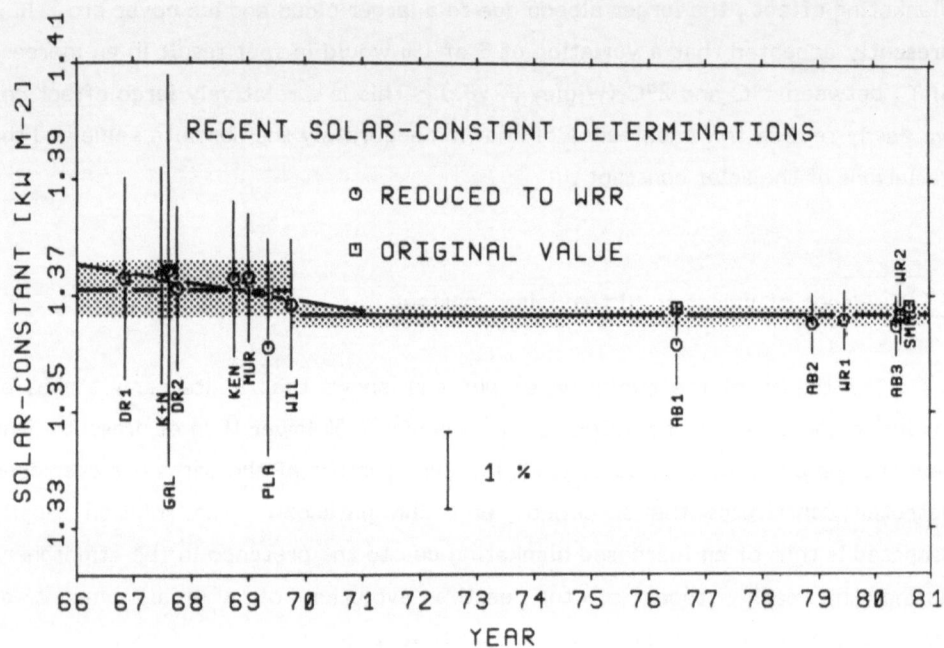

Figure 8   Summary of the determination of the solar constant from 1966 to 1980 . The straight portion of this curve corresponds to the period when absolute radiometers of modern design were used . The more precise measurements have been made only during the past three years on board the Nimbus 7 and the SMM satellites (Fröhlich and Brusa , 1981) .

---

appearance and with the size of sunspots at the surface of the sun .   More interesting , both the Nimbus 7 and the SMM data reveal a decrease of the solar constant of the order of 0.03-0.05% per year over the past two years .   These variations should be compared with the recent earth temperature trends .

TABLE IV

CHARACTERISTIC TIMES OF THE VARIOUS COMPONENTS OF THE EARTH

Atmosphere	10-100 days
Hydrosphere	1-1000 years
Cryosphere	1-1 million years
Biosphere	100-1 billion years

5.2 Evidence of the Temperature Variations of the Earth

What evidence do we have that the earth's surface temperature varies ? First of all we must clearly point out that we are only concerned here with time scales of several tens of years , i.e. time scales that are large with respect to meteorological fluctuations . As indicated in Table IV , there are several orders of magnitude between the characteristic time scales of the atmosphere and those of the hydrosphere , the cryosphere or the litosphere . In order to obtain statistically meaningful data for any climatic effect to be derived , averages over at least 30 years must be considered . Figure 9 (Morel , 1979) shows the result of averaged measurements of T in the northern hemisphere made from 1880 to 1970 which indicate a tendency of a gradual increase until the late 1930's corresponding to $\simeq 1°C$/century . This was followed by a symmetrical cooling which apparently is continuing .

5.3 Discussion

How much of this variation can be attributed to variations in the solar constant ? It is presently difficult to say . It would correspond to a decrease of 0.01%/year of the solar constant . That is not incompatible with the trends

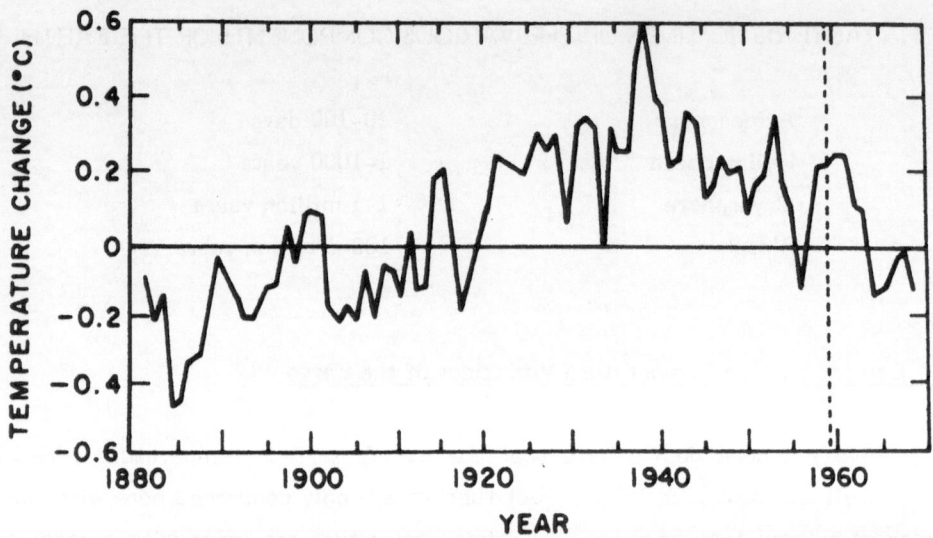

Figure 9   Running averages over 30 years of the temperature of the northern hemisphere showing a gradual increase until 1940 corresponding to a mean variation of +1°C per century then followed by a symmetrical decrease (from Morel , 1979) .

already noticed in the Nimbus 7 and SMM data . However , as we shall see in a moment , several other facts that certainly play a role must be taken into account and they tend to make our comparison fuzzier .

What evidence do we have of temperature variations over larger time scales ? Here we lack direct measurements . Several methods can be used such as the analysis of rocks , of tree rings (dendrochronology) , or of sediments in lakes and oceans , etc.. . Table V summarizes the various methods that are presently being used and indicates the time range to which they correspond .

TABLE V

Source	Measurements	Geographical domain	Temporal range (years)	Climatic information
Trees	Tree rings (dendrochronology)	Continents	1 000	Temperature Rain
Lakes (sediments)	Pollen	Continents	10 000	Temperature Rain
Polar caps	$^{18}O/^{16}O$ D/H	Greenland Antarctica	100 000	Temperature
Deep-sea sediments	Plancton	Oceans	1 000 000	Temperature
Rocks & sediments	Fossiles minerals	Global	1 million to 1 billion	Temperature CO_2

We find in Wigley (1981) a smoothed picture of the high and low latitude temperature trends over the past 100 million years. Figure 10 from Berger (1981) shows the climatic variations over the past 400 000 years and show evidence of strong fluctuations at a scale of the 100,000 years level.

Can we be sure that all these variations are due to the sun ?

Figure 11 lists the various competing processes that may potentially change our climate and implies that the answer to our simple question may in effect not be so simple. We see also from this figure that some effects have

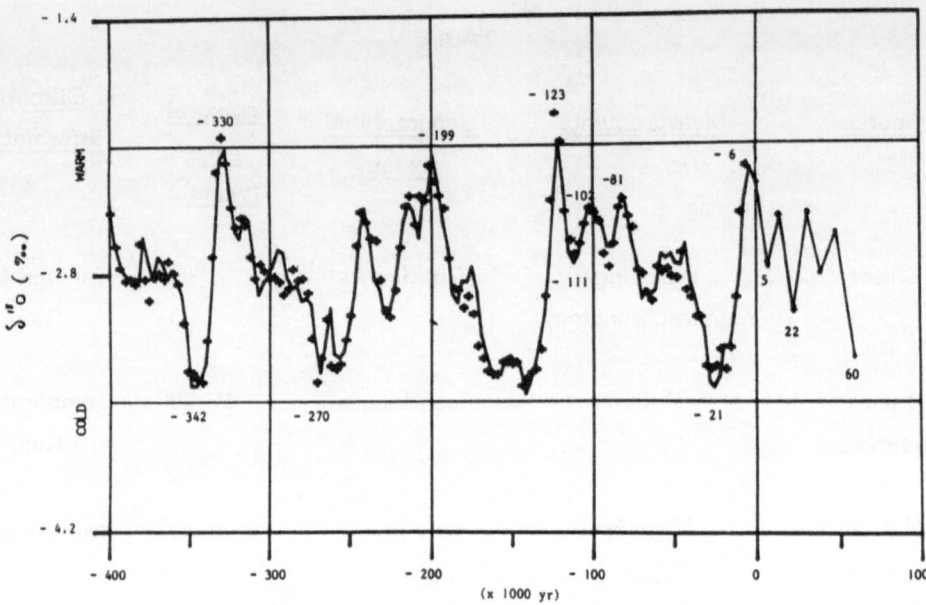

Figure 10   Long term climatic variations over the past 400,000 years . Crosses represent the abundance of Oxygen 18 (an isotope of oxygen) which is nearly a direct measure of temperature , as deduced from deep sea cores measurements . The full line is a model computed on the basis of the Milankhovitch theory , taking also into account the various positive as well as a negative feed back effects resulting from blanketing in the earth's atmosphere , ice caps coverage etc...The agreement between the theory and the measurements is very good . It allows us to predict the variation of the climate for the next 60,000 years and the possible occurence of a future major glaciation (from Berger , 1980).

different time scales and are more important at different epochs . The whole problem of **climatology** (past , present and future) bears on the correct evaluation of all these processes . For example , solar variability (about which we are concerned here) and volcanic activity may easily have competing effects on time scales of 1 to 10 million years .

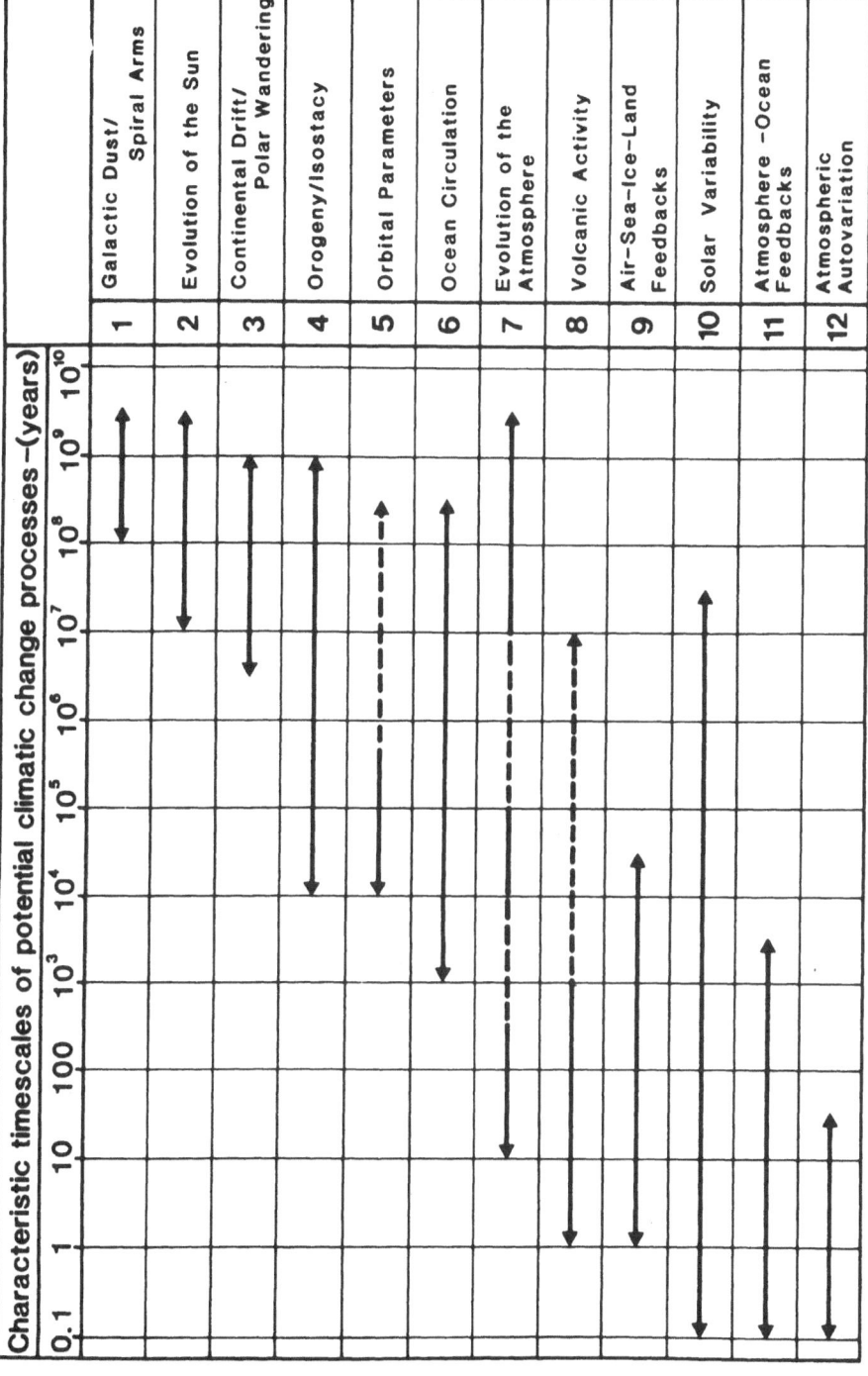

Figure 11

Presently it is honest to state that **we have no evidence that solar variability has any detectable influence on our climate** . This does not mean that this influence does not exist : we are just not able to extract it from that of the other processes , some of which have unambiguous effects . For example , we have more and more confirmation that the long term variation in the orbital parameters of the earth , such as the eccentricity of its orbit , the precession and the change in the tilt (obliquity) of its axis of rotation relative to the plane of the ecliptic , may explain most of the fluctuations of the earth's climate in the past million years or so (Berger , 1981 , Wigley , 1981 , Morel , 1979) . This is the well known **Milankhovitch** theory of climate forcing by orbital parameters.

Further back in the history of the earth , other phenomena must be considered , in particular the distribution of land over the globe . As shown on Figure 10 , the Milankhovitch theory remodelled by Berger , who took into account also terrestrial phenomena such as ice and sea coverage , not only succeeds in reproducing the past evolution of the earth's climate but is also able to predict its future trends. Assuming no human interference , it is predicted that the general cooling that started 6,000 years ago will continue over the next 25,000 years at least . A major glaciation is predicted 60,000 years from now . On the other hand , 10,000 years ago when the earth received 7% more solar radiation in June than at present , the atmosphere was warmer everywhere ; the arctic surface temperature was 6°C higher , and 4°C higher temperature prevailed at 30°N .

Another major process indicated in Figure 11 is that of the evolution of the earth's atmosphere . Changes in the atmospheric concentration of radiatively active gases such as CO_2 , H_2O and NH_3 can affect the **greenhouse effect** very otrongly . In particular the trend of increased CO_2 concentration , resulting from an increasing industrial activity and from fossil fuel , possibly constitutes one of the most important factors of systematic climatic change in the future, not only because of the greenhouse effect of CO_2 but also because of its effect on stratospheric ozone . Other species such as chlorofluoromethanes also affect the climate through the balance of ozone (Mitra , 1981) . In fact a number of climatic effects occur through ozone . This phenomenon is complex and not well understood . We refer the reader to a review of the coupling processes , by Ramanathan (1980) . These processes occur through mixing induced by planetary waves that dynamically couple the stratosphere to the troposphere .

6. Conclusion

It is now time to conclude . The purpose of this paper was to review what kinds of influence the sun may have on our immediate environment , on our daily life and on our climate .

Some phenomena , directly connected to solar activity , like the precipitation of atomic particles from the sun , have a strong and immediate impact . They are , for example , responsible for power black out on the earth in the vicinity of auroral zones . However , as disturbing as they may be , these effects are not of the highest concern to us : humanity has been living with them for many years without too much inconvenience and solutions can be found easily to the problem .

The consequences of the interaction of radiation with the atmosphere and of its reaction to solar fluctuations are much more subtle to evaluate . The closer we are to the earth's surface , the larger and the more complex are the various processes involved . Furthermore , the accuracy of the measurements , of the solar flux in the ultraviolet in particular , is far from reaching the value required to investigate more precisely the importance of these processes . However , remarkable progress has taken place in the past years and we are now in a much better position , in particular in the area of the measurements of the solar constant .

We now know more precisely that solar phenomena are only some among several causes of climatic variations . We must be concerned with the fact that man does not really know precisely what he is doing to his immediate environment . In particular , the release of nitrogeneous and chlorine compounds may have irreversible effects on the atmosphere of the earth through the chemistry of ozone . Hopefully we are now realizing the danger . It is time to evaluate with the greatest seriousness , with all possible means and with the best obtainable accuracy the influence of all the processes involved in the thermal as well as in the chemical balance of the earth and of its atmosphere . In the end we may then be able to control our own damage . May I just say at this time that the solar influences , considered separately , are probably those that present the smallest dangers to us .

Among other peaceful uses of space one will be to provide the means of continuously monitoring the solar output (particles and radiation) . The accurate knowledge of it can only be obtained from space . It is only through the organisation of an international endeavour that we will succeed in this ambitious project of paramount importance and hence to survive and maintain the earth as it should be .

References

Baum , W.A. , Johnson , F.S. , Oberly , J.J. , Rockwood , Strain , C.V. , and Tousey , R. , 1946 , Phys. Rev. 70 , 81

Berger , A. 1980 , "Sun and Climate" , Proceedings , Toulouse , 30 September-3 October , 1980 , CNES , CNRS DGRST , Ed. , 325

Fröhlich , C. and Brusa , R.W. , 1981 , "Physics of Solar Variations" , Proceedings , 14th ESLAB Symposium , V. Domingo Ed. , D. Reidel Publishing Company , 209

Keating , G.M. , 1981 , "Physics of Solar Variations" , Proceedings , 14th ESLAB Symposium , V. Domingo Ed. , D. Reidel Publishing Company , 321

Kockarts , G. , 1981 , "Physics of Solar Variations" , Proceedings 14th ESLAB , Symposium , V. Domingo Ed. , D. Reidel Publishing Company , 295

Mitra , A.P. , 1980 , "Sun and Climate" , Proceedings , Toulouse , 30 September -3 October 1980 , CNES , CNRS , DGRST , Ed. 121

Morel , P. , 1979 , "Le Climat" , Compte Rendu de l'Académie des Sciences , séance du 26 février 1979

Ramanathan , V. , 1980 , "Climatic Effect of Ozone Change" , Low Latitude Aeronomical Processes , Ed. , Mitra A.P. , Pergamon Press , 223 .

Schwentek , H. and Elling , W. , 1981 , "Physics of Solar Variations" , Proceedings 14th ESLAB Symposium , V. Domingo , Ed. , D. Reidel Publishing Company , 355

Svalgaard , L. and Wilcox , J.M. , 1978 , Ann. Rev. , Astron. Astrophys. , $\underline{16}$, 429

Wigley , T.M.L. , 1981 , "Physics of Solar Variations" , Proceedings 14th ESLAB Symposium , V. Domingo Ed. , D. Reidel Publishing Company , 435

Wilcox , J.M. , 1980 , "Sun and Climate" , Proceedings , Toulouse , 30 September-3 October , 1980 , CNES , CNRS , DGRST , Ed. , 173

Wilcox , J.M. and Scherrer , P.H. , 1981 , "Physics of Solar Variations" , Proceedings 14th ESLAB Symposium , V. Domingo Ed. , D. Reidel Publishing Company , 421

CHAPTER III

THE SIZE , SHAPE AND TEMPERATURE OF THE STARS

by R. Hanbury Brown

Chatterton Astronomy Department
School of Physics , University of Sydney , NSW 2006
Australia

1 . Introduction

People have always looked up at the night sky and wondered what the stars are and how they came to be there . There are many beautiful myths and legends which claim to give us the answers , but it is only in the last 100 years that the science of Astronomy has begun to tell us what the stars are really like . In this talk I am going to discuss three specific questions , how **hot** and how **large** are the stars and what is their **shape** ? Furthermore , I ought to make it clear that I am going to discuss these questions largely from my own point of view . The answer to our second question , how did the stars come to be there , belongs to Professor Longair's talk on Cosmology .

Fortunately it turns out that we can answer these three questions without actually visiting the stars with a tape measure and a thermometer ; it can all be done from the Earth . Firstly we must measure how large the stars appear to be in the sky or , in technical terms , their <u>angular size</u> . As we can all see , the angle which the Sun subtends to our eyes is roughly half a degree ; what we need to know is the corresponding angle for a more distant star , Sirius for example . Secondly we must measure the distance of the stars by observing how their position on a map of the Sky changes as the Earth goes around the Sun or , as

73

R. M. West (ed.), Understanding the Universe, 73–92.
© *1983 by D. Reidel Publishing Company.*

astronomers say , their <u>parallax</u> . Thirdly we must measure how bright the stars appear to be or in technical terms , their <u>light flux</u> . Now by multiplying our measurement of angular size by the distance of the star we can , be elementary geometry , find the actual **physical size** of a star in kilometres . By dividing our measurement of light flux by the square of the angular size we can find the light flux at the surface of the star and hence its **temperature** . Finally by making measurements of angular size in different orientations , we can find out something about the **shape** of the star .

All these things , size , temperature and shape are vital to our general understanding of what a star is really like ; they help us to explain why one star is different from another , what holds it together , where its energy comes from , how it evolves and so on .

If , by-the-way , you are wondering whether all this knowledge is worth while , may I remind you that an understanding of stars is important not only to Astronomy but to the progress of Science as a whole . The sky is a vast laboratory in which we study the behaviour of matter under extremes of temperature and pressure and on a scale which we cannot reproduce on Earth. Let me illustrate this with a few examples . It was by looking at the motions of the planets , and particularly at the moon , that we came to understand the basic laws of motion and gravity on Earth . It was by looking at the stars that we first discovered thermonuclear energy , and it was by thinking about the stars that we came to understand an important chapter in the story of Evolution - the evolution of the elements . Many people have heard about the evolution of Man from a primeval soup on Earth , but very few know that the heavy elements evolved from primeval hydrogen in the stars .

2 . History

For most of history our knowledge of the stars has been limited to measuring their positions on a map of the sky and to making crude estimates of their brightness . Before the invention of the telescope and the photographic plate these measurements were made by the naked eye using simple mechanical aids. At first people used arrangements of standing stones such as at Stonehenge and Carnac , and later they evolved sophisticated mechanical instruments such as those used by **Tycho Brahe** at Uraniborg in the 16th century . The last great observatories to be built without telescopes were erected in India by **Jai Singh** in the early 18th century ; interestingly they were built roughly a 100 years after the introduction of the telescope into astronomy and 50 years after the foundation of the Royal Observatory at Greenwich .

The introduction of the telescope in the early 17th century made it possible to see fainter stars and to make more precise measurements of their position , which was useful for navigation but told us very little about the stars themselves . It was not until the spectroscope was introduced in the mid-19th century that we got our first information about what the stars were really like . The spectroscope gave us the colour - more precisely the **spectral distribution** - of their light and , in so doing , told us about their chemical composition and gave us a rough idea of their surface temperature . Before the spectroscope we knew that stars made pretty patterns on the sky , afterwards we began to find out what they are made of and how hot they are .

But what about their size and shape ? To find the size of a star we must measure , as I have just said , both its angular size and distance . Let us look back for a moment at some of the early attempts to do this . In the 16th century Tycho Brahe reached the conclusion , by looking at the stars by eye , that a first magnitude star presents a circular disc with a diameter of 120 seconds of arc and that a fifth magnitude star , roughly the faintest star you can see , has a disc of 30 seconds of arc . In the early 17th century **Galileo** turned his telescope onto the stars and found that Tycho was quite wrong ; the stars , even when "magnified" by the telescope , looked like points of light . This was not , may I remind you , a matter of idle curiosity , it had an important bearing on one of the trickiest questions of the day , whether or not the orthodox Ptolemaic cosmology , as taught by the Church , was true .

If , as **Copernicus** had suggested , the Earth travels around the Sun , then it was to be expected that the apparent position of the stars , at least of the nearest

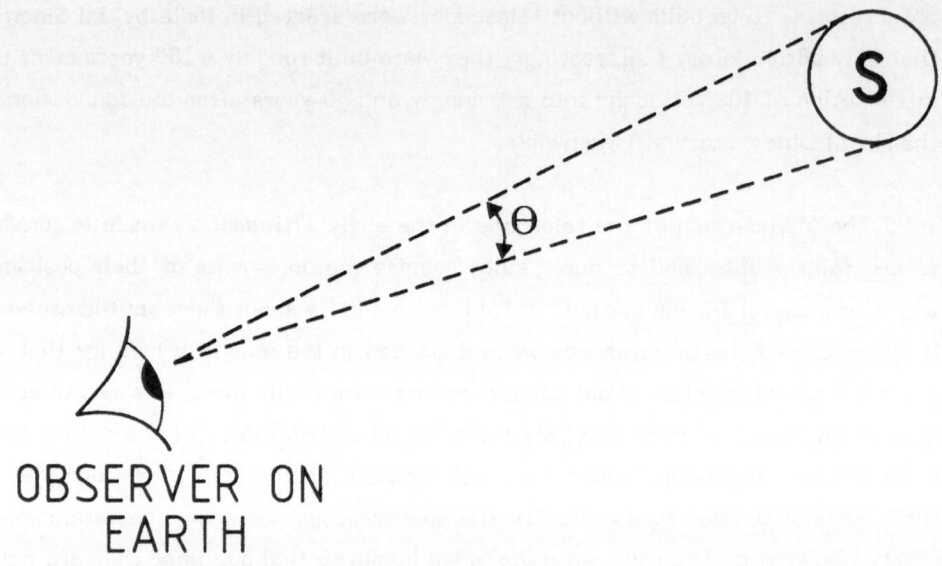

OBSERVER ON
EARTH

Figure 1 The angular diameter (Θ) of a star .

---

stars , would show an <u>annual variation</u> or <u>parallax</u> . Careful observations by Tycho
Brahe for example had shown that the stars did not appear to move annually or ,
to be more exact , that any such movement must be less than 1 minute of arc .
This limit set a minimum possible distance to the stars and , combined with
Tycho's estimates of their angular size , implied that if the Earth really does go
around the Sun then the stars must be bodies of incredible size , as large as the
whole orbit of the Earth around the Sun . Very few people were prepared to
believe that the stars were so large ; it is not surprising that Tycho himself
refused to accept the Copernican system .

But to Galileo the discovery that the stars looked like points of light , even through a telescope , was of great importance , because it made the Copernican theory more plausible . Indeed he tried to measure the angular size of the bright star Vega by a different method . He hung a fine silk cord vertically and then measured the distance from the cord at which he had to stand so that it just occulted the star . By careful experiment he reached the conclusion that the angular size of Vega was 5 seconds of arc . About 350 years later we measured Vega again at Narrabri Observatory in Australia ; Galileo's result was roughly 1500 times too large . Nevertheless Galileo's measurement served his purpose , it showed that Tycho's estimate of the angular size of a star was much too big , and gave a plausible explanation for the failure of astronomers to observe parallax; quite simply the stars are much too far away .

The first successful measurements of the distance of a star were made in 1838 , by which time astronomers had developed sufficiently accurate methods of measuring the relative positions of stars seen in a telescope . One of the first measurements was made by **Bessel** working at Königsberg ; he found the parallax of 61 Cygni to be about 0.3 seconds of arc , which corresponds , very roughly , to a distance of 10 light-years or 100 million million kilometers . Since those days the accuracy of measurements has greatly improved , for example by photography , and it is now possible to measure the parallax of a star with an uncertainty of about 0.01 seconds of arc . As a consequence we know the distance of several hundred of the nearer stars with an uncertainty of less than 10 per cent and of several thousands of stars with a somewhat greater uncertainty .

The first successful measurement of the angular size of a star had to wait for the development of a new type of instrument , a stellar interferometer , because it could not be done with a conventional telescope . In 1920 , at Mt . Wilson in California , **Michelson** and **Pease** used an interferometer to measure the angular size of the red supergiant star Betelgeuse (α Orionis) . Betelgeuse is an irregular variable and its angular size was found to vary between 0.034 and 0.047 seconds of arc . This result , when combined with the measurements of distance and light flux , showed an astonished public that the size of Betelgeuse varies between 200 and 300 times the size of the Sun and that its surface temperature is about 3000 K .

The smallest angle which could be measured with Michelson's stellar interferometer was about 0.02 seconds of arc and , in consequence , it was limited to measuring a total of 6 stars . All these stars are giants or supergiants and are by no means typical of the general stellar population . In an effort to measure smaller angles , and thereby to extend the measurements to more common stars , a larger interferometer was built at Mt. Wilson , but it failed to work satisfactorily and the programme was discontinued in the 1930's . Since that time there have been other attempts to develop Michelson's interferometer but , so far , they have failed . Indeed the technical problem of measuring the angular size of the common stars has proved to be much more difficult than measuring their distance and light flux . I shall therefore devote most of my time to reviewing these difficulties and to telling you what has been done , and might perhaps be done in the future , to overcome them . First of all let us take a quick look at what it is that we aim to measure , the angular size of an average star .

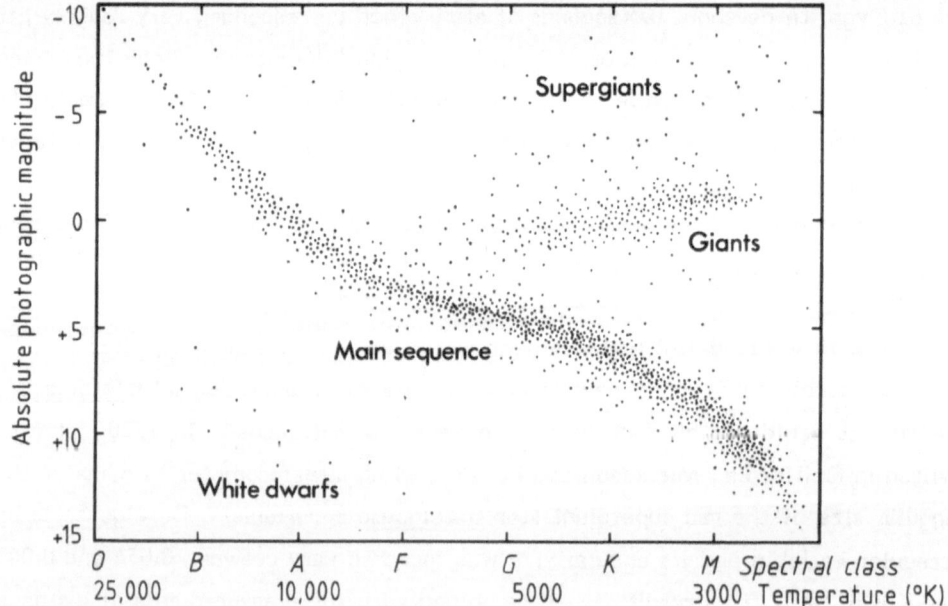

Figure 2 The Hertzsprung-Russell diagram for stars of known distance .

3. The Different Types of Star

We know from looking at their spectra that stars differ greatly in temperature and in the amount of light which they emit (luminosity). They are usually classified by their spectral type which, for historical reasons, follows a curious sequence OBAFGKM which corresponds to their surface temperatures. Thus type O stars are hot, blue and large, and type M stars are cool, red and small. If we plot luminosity against spectral type or temperature we get a remarkable pattern known as a Hertzsprung-Russell Diagram or H-R diagram for short. On this diagram the vast majority of stars lie in a narrow band which we call the main sequence and they are what I mean by common stars.

Main sequence stars shine by converting hydrogen into helium and their position on the sequence is determined by their mass; heavy stars are at the hot end and light stars at the cool end. In due course, when they have exhausted a substantial fraction of their hydrogen, all sorts of strange things happen. They leave the main sequence and become giants or supergiants, enormously large stars with distended and tenuous envelopes. What happens to them in the end depends largely upon their mass; they may blow up as a supernova, evolve into a white dwarf or a neutron star or, perhaps, form one of those darlings of modern astrophysics, a black hole. But we haven't time for all that. The point I want to make is that the vast majority of stars, whose structure and evolution we claim to understand, lie on the main sequence; they are the common stars.

If now we want to measure the angular size of common stars theory tells us that we must be prepared to measure extremely small angles, much smaller than the 0.02 seconds of arc which was achieved by Michelson's interferometer. The actual range of angles will depend on how large a sample of the main sequence we wish to make. Contrary to what one might expect, it is the hot stars which are the most difficult; although they are physically the largest stars on the main sequence, the nearest specimens are so far from the Earth that they present a very small angular size. In fact if we want to collect a small sample of main sequence stars, including half a dozen of the hottest stars (type O), we must aim to measure the absurdly small angle of about 5×10^{-5} seconds of arc , which is roughly the apparent angular size of a tennis ball on the moon .

No wonder it has proved difficult to measure the common stars !

4 . The difficulties of developing Michelson's stellar interferometer

A simplified outline of Michelson's stellar interferometer is shown in Figure 3. Light from a star is received on two small movable mirrors M_1 , M_2 and

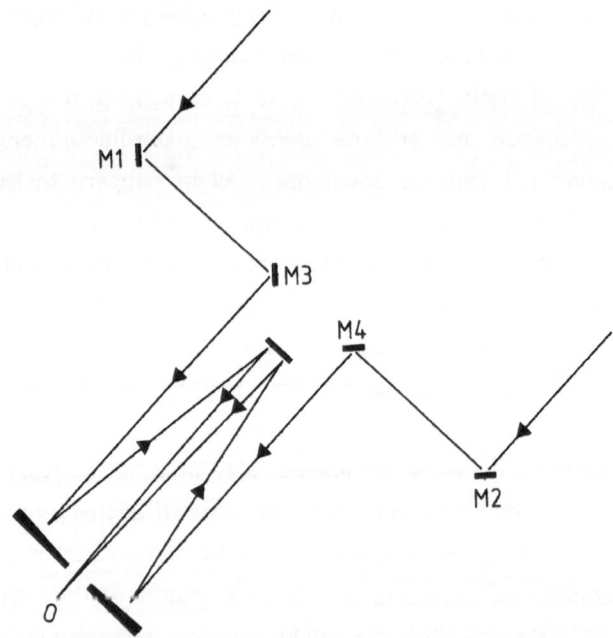

Figure 3 Outline of a Michelson Interferometer .

is reflected to the fixed mirrors M_1 , M_2 which in turn reflect it to the focus of the telescope at O . The mirrors M_1 , M_2 are mounted on a beam , and their separation could be altered at will by the observer . The two images of the star are superimposed in the focal plane and , when the instrument is in proper adjustment , the combined image is crossed by alternate bright and dark bands called fringes. The contrast or "visibility" of these fringes is a measure of the similarity , of the light beams received at the two spaced mirrors . When the two mirrors are very close together they both see the same "view" of the star and the fringes are very clearly visible . As the two mirrors are separated they see different "views" of the star and the visibility of the fringes decreases until they disappear . Roughly speaking the fringes disappear when the two mirrors are separated by a distance λ / Θ , where λ is the wavelength of the light and Θ is the angular diameter of the star . Thus when Pease first observed Betelgeuse on 13 December 1920 the fringes disappeared at a mirror separation of 3.05 m giving the angular size of the star as 0.047 seconds of arc .

Michelson's stellar interferometer was mounted on the 100 inch telescope at Mt. Wilson and the maximum possible separation between the two small mirrors (M_1 , M_2) was only 6m . For this reason the smallest angle which it could measure was 0.02 seconds of arc . As we have just seen , to observe only a small sample of main sequence stars we must aim to measure the very much smaller angle 5×10^{-5} seconds of arc and , to do that with Michelson's interferometer , we should have to increase its size from 6m to more than 1km . For many reasons this cannot be done simply by making a larger version of the original instrument . Let us look briefly at two of the main reasons why this is so .

The first and most obvious difficulty which we face in any attempt to build a large instrument is the need for extreme mechanical precision . In Michelson's interferometer the two beams of light from the separated mirrors must be brought together at the focus through paths which are very closely equal in length . For example , if we choose to look at the fringes by eye then we must build the instrument with sufficient rigidity , and point it at the star with sufficient precision , to ensure that the two paths are equal to about 1 wavelength of light or about 0.0005 mm . That , I need hardly point out , is a very tight mechanical tolerance for a large instrument !

The second major difficulty is the effect of the Earth's atmosphere . Why , we may ask , did Michelson mount an interferometer on a telescope of comparable size ? Why didn't he look directly at Betelgeuse through the telescope ? After all , as Airy showed in 1835 , the angular resolving power of a telescope is the same as that of an interferometer of the same overall size . Unfortunately this is only true in space . On Earth the light from the stars is randomly bent and delayed by turbulence in the atmosphere . The characteristic size of the turbulent elements is about 10 cm and they move with the wind . Because these elements differ in refractive index they introduce random patches of phase and amplitude into the starlight falling on the telescope and , roughly speaking , limit its resolving power to that of a telescope with the diameter of a single patch , roughly 10 cm . As a consequence the actual image of a star as seen through a large telescope depends upon the site and the weather and is seldom less than 1 second of arc . In other words , large telescopes have been built mainly because they collect more light and not because they show us the size of the stars ; for most of the time their angular resolving power is not much better than that of a small telescope . Thus , if we were to look at Betelgeuse through the 100 inch telescope we should see a fuzzy disc roughly 1 second of arc across ; we should certainly not see a star with an angular size of 0.047 seconds of arc .

The success of Michelson's interferometer showed that it was not so badly upset by the atmosphere as the 100-inch telescope on which it was mounted; nevertheless , it was by no means immune . As far as we can see it worked because the two little separated mirrors (M_1 , M_2 in the figure) were comparable in size with the 10 cm turbulent elements of the atmosphere and so the phase and amplitude of the light across them was reasonably uniform , at least for short periods of time . They were therefore able to form fringes , although as the patches of "seeing" drifted across the field these fringes moved . In fact for most of the time the fringes danced about in the focal plane and only when they were not moving too fast could the observer see and measure them . As one might expect , the observed visibility of the fringes depended upon the state of the atmosphere and , as a result , the accuracy of the measurements was unacceptably low by the standards demanded by modern astrophysics . The effects of the atmosphere are likely to prove even more serious in a larger instrument .

The two limitations , the effect of the atmosphere and the need for extreme mechanical precision , have so far prevented the development of Michelson's interferometer . A larger model with a mirror separation of 15m was in fact built at Mt. Wilson by Hale and Pease around 1930 . It proved extremely difficult to see the fringes with this instrument , let alone to form an estimate of their visibility , and despite several years work no reliable measurements of stars were ever made . It seems that the original 6m interferometer was near the limit of the technique at that time .

5 . A New Type of Stellar Interferometer - An Intensity Interferometer

The next step forward was made in 1950 when we developed a completely new type of instrument , an **intensity interferometer** , for use in radio-astronomy at Jodrell Bank (University of Manchester , U.K.) . At that time the major problem in radio-astronomy was to discover the nature of the discrete sources of radio waves which had recently been discovered . In those days we called them radio stars , but we didn't know whether they were really stars , nebulae or galaxies . One obvious way to find out was to measure their angular size and the simplest way of doing that was to make a radio analogue of Michelson's interferometer . However we knew that , if these mysterious sources should really prove to be stars , we would need an enormously long interferometer ; for example , to measure an angle of 0.01 seconds of arc at a wavelength of 1m , we would need a baseline between the two antennae of 20,000 km ! Nowadays it is technically possible to make such an instrument but in those days it would have been extremely difficult .

We solved this problem by developing an intensity interferometer which can be operated with very long baselines ; although , as it turned out , the so-called "radio stars" proved not to be stars and were resolved with baselines of only a few kilometres . Nevertheless our work was not wasted because we realized that , by using an intensity interferometer for light waves , we would solve the two main problems which had prevented the further development of Michelson's interferometer . We could now see how to build an instrument to measure the common stars .

As we have seen , one of the major difficulties in extending Michelson's interferometer is that the light waves from the two separated mirrors must be brought together without upsetting their relative phase and amplitude . This difficulty is completely avoided in an intensity interferometer . The light at the two separated mirrors is detected by photoelectric detectors , whose output currents are proportional to the intensity of the light , and these two currents are then brought together and compared in an electronic correlator . The intensity of the light wave from any hot object , such as a star , fluctuates rapidly because the wave itself is a random combination of many component waves . For this reason the output current of a photoelectric detector contains fluctuations which correspond to the fluctuations in the intensity of the light itself . When two detectors are close together , and see much the same "view" of a star , the fluctuations are the same in both detectors and their correlation is therefore high . As the two detectors are separated this correlation decreases , much in the same way as the fringe visibility decreases in Michelson's interferometer . Thus by measuring the correlation at different baselines an intensity interferometer can be used to measure the angular diameter of a star .

An intensity interferometer solves both of the principal difficulties with Michelson's interferometer . It can easily be made large enough to measure common stars and it is almost completely unaffected by turbulence in the Earth's atmosphere . The first of these advantages is due to the fact that it is technically much simpler to bring two currents together than two light beams . For example , if in an intensity interferometer we restrict the frequency range of the fluctuations which we correlate to 0-100 MHz , then we need only ensure that any differences between the two paths from the star to the correlator are less than a few centimetres ; clearly , this is mechanically much easier , roughly a thousand times easier , than building an enlarged version of Michelson's interferometer in which the tolerances are measured in microns . The effects of the Earth's atmosphere on an intensity interferometer are reduced for much the same reason . Although the relative amplitude and phase of the light received from a star at two separated points is seriously disturbed by the Earth's atmosphere , the "low-frequency" fluctuations in intensity are not , and the information which they bring about the angular size of the star , surprisingly , survives .

But there is , I regret to say , a price to be paid for these great advantages . Quite simply , an intensity interferometer is insensitive and demands an enormous lot of light . As we shall now see , to reach even bright stars one must build very large light collectors .

6 . The Stellar Intensity interferometer at Narrabri Observatory

We built our first intensity interferometer at Jodrell Bank in 1955 with the mirrors of two large military searchlights , and used it successfully to measure the angular diameter of Sirius . This was the first measurement ever to be made of a main sequence or common star . The observations were all carried out under conditions of bad "seeing" ; Sirius twinkled violently the whole time and there was no doubt that the equipment would work through a turbulent atmosphere . The next step was to build a full-scale instrument which , in due course , was installed at Narrabri Observatory (Australia) in 1962 .

The stellar interferometer at Narrabri was a very unusual instrument . Two very large optical reflectors (6.5 m in diameter) were mounted on a circular railway track 188m in diameter . At the focus of each reflector there was a photomultiplier on which the reflector focussed the starlight . The output currents from the two photomultipliers were carried by coaxial cables to a central laboratory where an electronic correlator measured their correlation in the frequency band 0-100 MHz . In operation the two reflectors followed the star in azimuth by moving around the circular track and in elevation by tilting about a horizontal axis . The separation between the reflectors could be varied from 10m to 188m and they always moved so that the line joining them was at right angles to the direction of the star .

In the course of a seven-year programme we measured the angular sizes of a sample of 32 single stars in the spectral range type O to type F ; many of these stars are common stars . The size of our sample was limited by the low sensitivity of the equipment which required very long exposure times ; for example , to measure a star of magnitude +2.5 with an accuracy of $\pm$ 5 per cent required an exposure of 100 hours which , quite often , took us over a month to

Figure 4 The two reflectors of the Stellar Intensity Interferometer at Narrabri Observatory (N.S.W. Australia).

get .   Our sample was not limited by the angular resolving power of the instrument ; with a maximum separation of 188 m between the two reflectors we could measure an angle of 2×10^{-4} seconds of arc which was more than adequate for any star bright enough to be measured in a reasonable time .

For these 32 stars the angular sizes were combined with photometric measures of light flux to find their surface temperatures or , to be more exact , their effective temperatures . Both the photometry and the analysis were carried out by a combined team from the Universities of Sydney and Wisconsin. Most of the photometry was done on the ground by conventional methods , but for the hotter stars this was impossible because much of their light is ultra-violet and does not reach the Earth ; it is absorbed by the ozone of the Earth's atmosphere . For the hot stars the measurements of light flux were made by the **Orbiting Astronomical Observatory (OAO-2)** .   The results established a new scale of effective temperature for stars in the spectral range Type O to Type F . At one end of the scale is the Type O star Naos (ζ Puppis) with an angular diameter of about 4×10^{-4} seconds of arc and a temperature of 32,500 deg. K ; at the other end is the Type F star Procyon (α CMi) with an angular diameter of about 5.5×10^{-3} seconds of arc and a temperature of 6500 deg. K . This is the first temperature scale for hot stars which , apart from some minor corrections , is based entirely on observations ; previous scales have been made by comparing the observed spectra of stars with theoretical spectra at different assumed temperatures . The temperature scale is an essential link between the observed properties of stars and the theories of stellar structure and atmospheres . It could not have been established , I should point out , without the help of the Orbiting Astronomical Observatory .

The angular diameters which we measured at Narrabri have also been used to find the sizes of different types of stars , that is to say their actual physical diameters in kilometres . Rather surprisingly we could do this for only 15 of the 32 stars as they are the only ones whose distances are known with reasonable accuracy ($\pm$ 25 per cent) ; the remaining stars are too far away for their parallax to be measured satisfactorily by existing techniques . Among these 15 stars there are 9 main sequence or common stars whose physical sizes ranged from 7.9 solar diameters for the hot type B1 star Spica (α Vir) to 1.6 solar diameters for the cooler type A3 star Fomalhaut (α PsA). The hottest star in the list, the type O5f star Naos (ζ Pup), which is almost, but not quite, a main sequence star was found to have a size of 15.6 solar diameters.

So much for the sizes and temperatures of the stars ; but what did the intensity interferometer tell us about their shape ? The first and most important fact that it told us about the "shape" of a star is whether it is **single** or **double**. Obviously we need to know this to interpret our observations of the brightness and spectrum ; furthermore the whole question of whether stars are single or binary has an important bearing on their evolution. There are several conventional ways in which binary stars can be detected but, even so, many of them escape detection. Indeed several of the well-known "single" bright stars which were observed at Narrabri proved to be binary. This result suggests that, if we could survey large numbers of stars with a more sensitive interferometer, we should greatly enlarge our knowledge of the statistics of multiple stars.

Several other observations were made at Narrabri on the "shape" of stars. In one of them we looked for the distortion in the shape of a rapidly spinning star. Theory predicts that the star should be **flattened** at the poles so that its angular size varies in different directions across the star and we actually observed this effect in the bright star Altair (α Aql). In another particularly interesting experiment we measured the angular diameter of the Wolf-Rayet star γ Velorum first in a very narrow band of wavelengths corresponding to a bright emission line in its spectrum and then in a relatively broad band of wavelengths outside this line. The bright emission line originates in an extensive cloud of gas surrounding a central hot star, whereas the light outside the line originates in the star. The observations showed that the apparent angular size of γ Velorum was larger by roughly five times when measured in the light of the emission line ; thus the measurements yielded the angular diameter and temperature both of the central star and of the surrounding emission region. This type of measurement, where the angular size of stars is measured in narrow spectral lines has, I believe, a considerable future which is, so far, almost unexplored. For example, such measurements would help us to understand the structure of stars with extended atmospheres and coronae and thereby throw light on the important topic of how stars lose their mass as they grow old .

An experiment which we would very much like to have done at Narrabri is to measure the changes in angular diameter of a **Cepheid variable star** as it pulsates . By combining these observations with spectroscopic measurements of the changes in radius , it should be possible to make a measurement of the distance to a Cepheid . Our scale of distance in the Universe is based , to a large extent , on Cepheids and an independent measurement of their distance would be valuable . Unfortunately the interferometer at Narrabri was not sufficiently sensitive and this attractive experiment remains one which we hope to do in the future .

We made many other observations which I haven't time to describe , but there is one which I must not omit . In an extensive series of experiments we showed that turbulence in the Earth's atmosphere , bad "seeing" , has no appreciable effect on an intensity interferometer .

7 . A Successor to the Narrabri Stellar Interferometer

Our work at Narrabri has shown what might be done with a more powerful instrument and , about ten years ago , we decided to build its successor . We planned to build a new interferometer which would be at least 100 times more sensitive than the original instrument and therefore reach stars of magnitude +7.5 . As a first step we choose to design a more powerful intensity interferometer for the obvious reason that we know it would work . But , to reach the sensitivity which we wanted , the instrument proved to be both very large and costly ; to build it in 1972 would have cost $4m . We decided to search for a more economical way of doing the job and , to cut a long story short , we chose to modernize Michelson's stellar interferometer . As far as we could see it should now be possible , by using the latest optical and mechanical techniques , to develop an instrument which is more sensitive and , at the same time , cheaper to build than an intensity interferometer . The major uncertainty is , of course , whether or not it can be made to work with the necessary accuracy through the Earth's atmosphere .

To test our ideas we have built a small pilot model which incorporates a variety of modern tecniques such as "active optics" , photon counting , very narrow optical bandwidths , laser-control of distances and so on . It is almost complete and should be ready in a few months time . And so , if all goes well , in a few years time we shall build a full-scale interferometer which will , we hope , measure the angular diameter of almost any type of star brighter than magnitude +7.5 . That will , I should add , allow us to choose from a population of about 20,000 stars .

8 . Lunar Occultation and Speckle Interferometry

I have so far talked mainly about our own work with the intensity interferometer ; nevertheless our group at the University of Sydney is not the only one which in recent years has been trying to measure the angular size of stars . There is , for example , a group in the U.S.A. which has been working on a technique called "lunar occultation" in which the light from a star is observed while it is being occulted by the edge of the Moon . This attractively simple method is particularly useful for detecting close-spaced binary stars and large cool stars .

Another group , now at C.E.R.G.A. in France , are developing a technique called "speckle interferometry" and have already shown how it can be used to remove the distorting effects of the Earth's atmosphere on the image of a star formed by a large telescope . The main application of this interesting technique has , so far , been to the study of close-spaced binary stars with large telescopes . However the group are working on its application to two-telescope interferometers for the measurement of faint stars and objects of very small angular size .

9 . Astronomy in Space

As we have seen astronomers have at last started to give some real answers to the ancient questions about the stars with which I started this talk . So far they have measured the size and temperature of only a few of the brightest stars and have told us a little about their shape . But that is only a start ; there is so much more to be done , especially by more powerful instruments and , of course , by working in space . Many of the experiments which we believe it will be possible to do , for example on Cepheid variables , emission-line and spinning stars , and on stars with coronae , are of great interest ; but maybe the ones which we cannot foresee will prove most interesting .

A major advance in observational astronomy is currently being made in space ; what bearing will it have on the questions we have been discussing ?

All the measurements about which I have been talking - light flux , parallax and angular size - are seriously limited by absorption or turbulence in the Earth's atmosphere and , at least in principle , can be done better in space . The measurements of light flux are the most seriously affected ; at one end of the spectrum the atmosphere absorbs the ultra-violet light from hot stars and , at the other end , it absorbs the infra-red radiation from very cool stars . However the technique and the facilities for photometry in space , such as the **International Ultra-violet Explorer** , have received a good deal of attention and a large number of stars have already been observed in the ultraviolet . The measurements of parallax have not yet migrated from Earth to space , but they are well on the way. We can look forward in the near future to a substantial improvement , by a factor of about 5 times , in the accuracy of parallax measurements when the satellite **Hipparcos** is launched . This should give us reasonably accurate distances to thousands of stars .

No doubt the measurements of angular size will , in due course , follow the others into space because one of the main obstacles to making them on Earth has been the effect of turbulence in the atmosphere . For example , if a Michelson-type interferometer were to be put into space it could be made much more sensitive than its ground-based counterpart because it could use much larger mirrors to collect the light , the mirrors would not be limited in size by the elements of atmospheric turbulence ; It could also be made to measure angular sizes in ultra-violet light . To exploit its greater sensitivity it would have to work with baselines of , perhaps , several kilometres . How that would be achieved with the necessary precision in space is a problem for the future , but is not , I would guess , insuperable .

From the strictly limited point of view of the questions with which we have been concerned in this talk , the value of an interferometer in space cannot be assessed until we know more about what can be done on the ground . We need to know , for example , how well a modernized version of Michelson's interferometer can be made to perform and what can be done with the two-telescope speckle interferometer being developed in France . All we can say at the moment is that interferometers on the ground are likely to be cheaper to develop , build and operate than in space !

But we must take a wider view ; the whole history of science , and particularly of Astronomy in recent years , tells us that progress depends on the development of new instruments which give us new ways of looking at the world . The Space Telescope , for example , will in a few years time tell us something new about the stars ; it will show us objects 100 times fainter than we can see at present and it will show them 10 times more clearly than our ground-based telescopes . Observations of X-rays , ultra-violet and infra-red will tell us new things about stars which are very hot and stars which are very cool . An interferometer in space might well tell us something new and exciting , for example , about the compact nuclei at the heart of quasars and Q.S.O.'s . Who knows what we shall find ? - it is always unwise to forecast what a new instrument will discover ; invariably the world turns out to be stranger than we think .

CHAPTER IV

X-RAY and GAMMA-RAY ASTRONOMY

by Professor Walter H.G. Lewin

Massachusetts Institute of Technology
Cambridge , Massachusetts 02139 , USA

1. Introduction

From the previous speakers , this morning , you have heard about research in astronomy which can be done better in outer space than from the ground . I will talk to you about research in astronomy that can only be done from outer space . The earth blanket is completely opaque for **X-rays** and **gamma-rays** ; none that are incident to the earth's atmosphere will reach the ground .

I will be talking about the same kind of X-rays that your dentist uses when he takes an X-ray picture . (For the physicists in my audience we are dealing here with an energy range , say , from 1-40 keV .) You are perhaps familiar with gamma rays through reading about radiation treatment of certain forms of cancer

* This text is a transcript (somewhat edited) of the talk as it was recorded .

R. M. West (ed.), Understanding the Universe, 93–127.
© 1983 by D. Reidel Publishing Company.

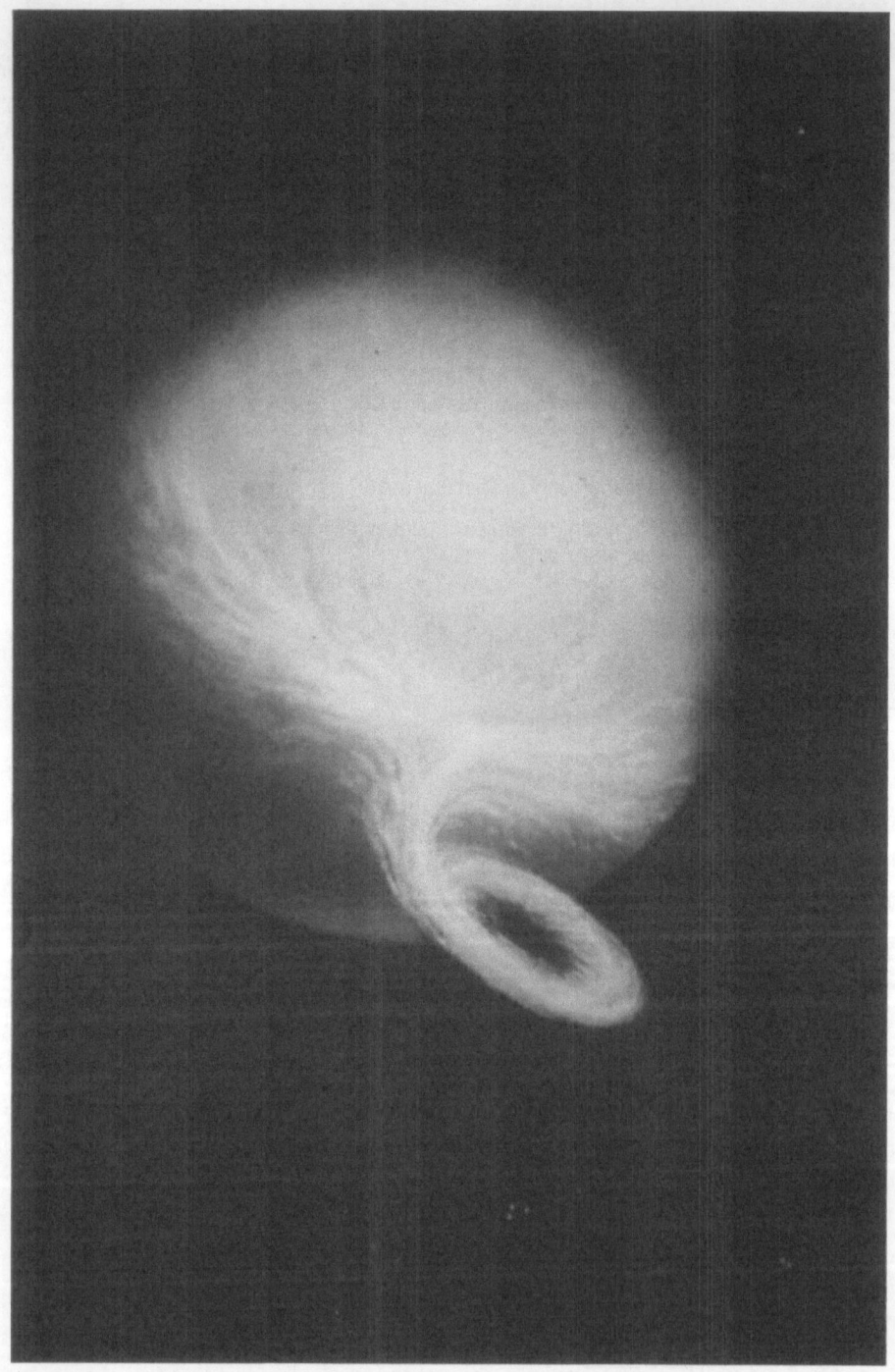

with cobalt 60 . Gamma rays are much more energetic than X-rays (typically one hundred to one hundred thousand times more) and in turn the X-rays are about ten thousand times more energetic than optical light , all being members of the family of electro-magnetic radiation .

2 . A Little Bit of History - The X-Ray Binaries - Neutron Stars - Black Holes

X-ray astronomy began in 1948 when German V2 rockets , which were leftovers from the unfortunate developments in World War II , were flown by U.S. scientists from the Naval Research Laboratory who discovered X-rays from the sun . Solar X-rays are but a modest byproduct of the sun ; almost all the radiation that we receive from the sun is in the visible part of the spectrum and in general only a minute fraction , less than one part in a million , is emitted as X-rays. However , objects were discovered in the sixties , from which a large fraction of the radiation (in many cases almost all the radiation) , is emitted in X-rays . If I also tell you that these objects produce a billion times more X-rays than our sun does , then you will realize that these objects must be very different from an ordinary star like our sun.

It was discovered in 1971 that these new "X-ray objects" were associated with **neutron stars** . Neutron stars were earlier discovered in 1967 when periodic radio pulses , separated by about a second , were discovered . It was soon

---

Figure 1   This is an illustration of an X-ray binary : a large "normal" nuclear burning star and an invisible compact object (a neutron star or a black hole) inside the ring-like structure . Gas flows from the "normal" star to the neutron star ; it heats up to tens of millions of degrees and produces predominantly X-rays . (Photograph courtesy of the National Geographic Society .)

understood that these radio pulses were the result of rotating neutron stars ;
objects which have radii of about ten kilometers but which contain an amount of
matter comparable to that in the sun . As a result , the density of the neutron
star's material is enormous and approaches that of nuclear matter . To give you
some feeling for how dense these neutron stars are , if you would take one big
tablespoon of the material from a neutron star you would have about the same
amount of matter than you have in all the buildings here in Vienna . The neutron
stars turned out to be crucial in the understanding of the new X-ray objects which
produce such tremendous amounts of X-rays.

Our concept of these enigmatic X-ray sources can be seen in Figure 1 . You
see a "normal" star , for instance one like that of our sun (it could be bigger or
smaller) and this star is in a **binary system** ; the companion star is a neutron star
(perhaps a black hole in a few cases) . The two stars are so close to each other
that gas is flowing from the "normal" star to the neutron star ; it is falling onto
the neutron star . Since the neutron star has an enormous gravitational pull near
its surface , the gas will reach enormous velocities . If you dropped a sugar cube
from far away onto the earth it would only achieve a modest velocity of about ten
kilometers per second when it enters the earth's atmosphere , but in the case of
the neutron star it would be speeding with a velocity of about 100,000 kilometers
per second (thus at about one third the speed of light) when it reaches the star's
surface . When the gas approaches and reaches the neutron star , it heats up
because of its high velocity and it will heat the neutron star surface to ten million
degrees and more . At such high temperatures matter radiates largely in X-rays
(the optical radiation produced is much smaller) .

To give you some feeling for the power of such neutron stars , imagine first
that we dropped a sugar cube unto a neutron star . The energy that would be
released when that sugar cube hits the surface would be approximately equivalent
to a hydrogen bomb explosion on earth ! Now imagine that 10^{16} sugar cubes fall
onto a neutron star every second (i.e. , about 10^{17} g/sec) ; the X-ray power is
then enormous indeed (about 10^{37} erg per sec) .

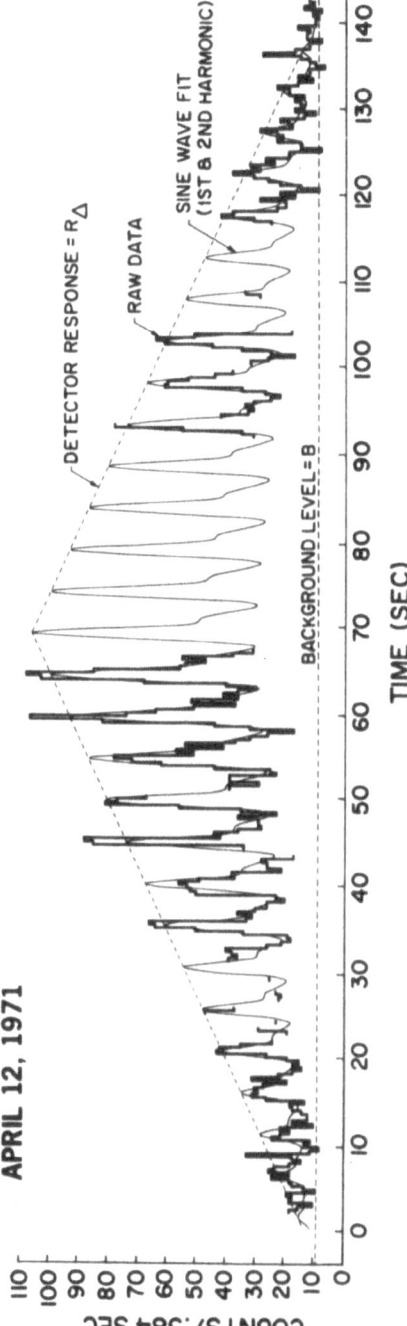

Figure 2 Discovery of X-ray pulsations from Cen X-3. The recorded X-rays (in counts per 0.384 sec) are shown for a 140 second portion of data obtained with the U.S. Observatory Uhuru when Cen X-3 drifted through its field of view. The pulses are due to the 4.8 sec rotation period of a neutron star in the binary system (see Figure 3). The triangular "envelope" is due to the object drifting in and out of the field of view of the X-ray detectors. (Courtesy of R. Giacconi, H. Gursky, E. Kellogg, E. Schreier, and H. Tananbaum, Astrophysical Journal Letters, 15 July 1971.)

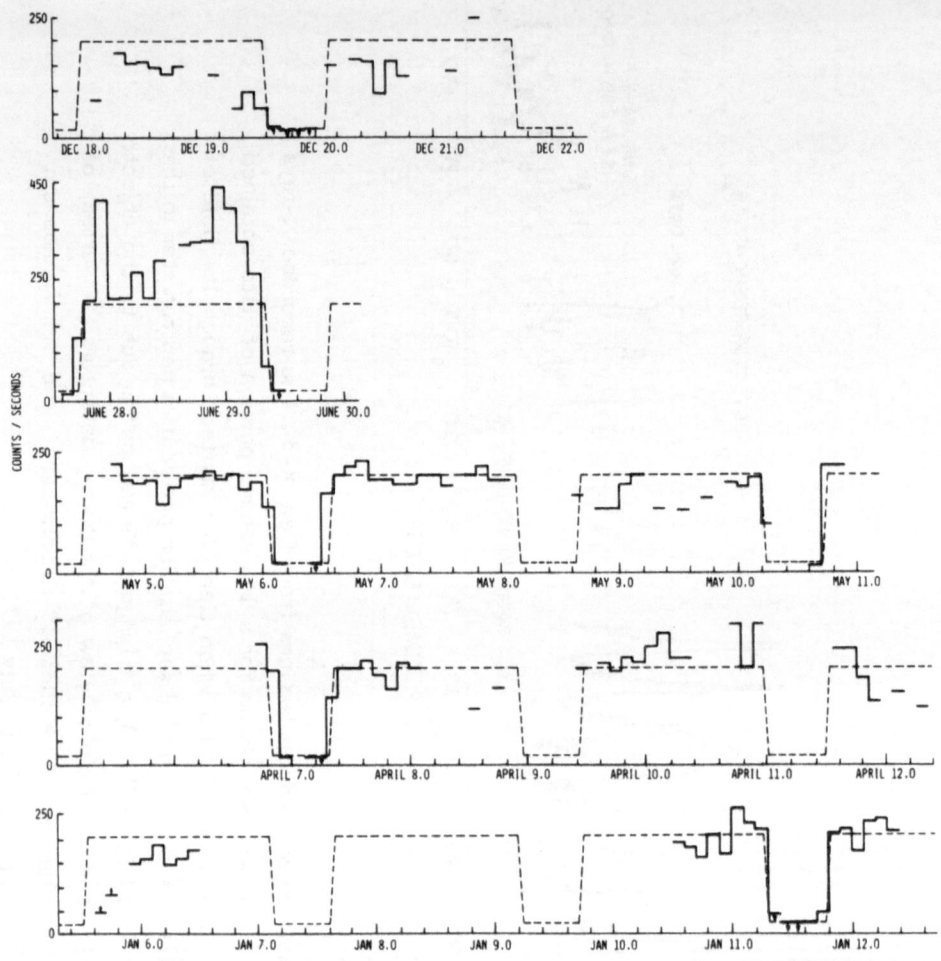

Figure 3   Discovery of X-ray eclipses from Cen X-3 . These are data taken with the Uhuru X-ray Observatory . The eclipses are the signature of a binary system of the kind shown in Figure 1 . The orbital period of the system is about 2.1 days ; it can be derived from the X-ray data . The eclipses occur when the neutron star (see Figure 2) "disappears" behind the companion star as seen from Earth . (Courtesy of E. Schreier , R. Levinson , H. Gursky , E. Kellogg , H. Tananbaum , and R. Giacconi , Astrophysical Journal Letters , 15 March 1972 .)

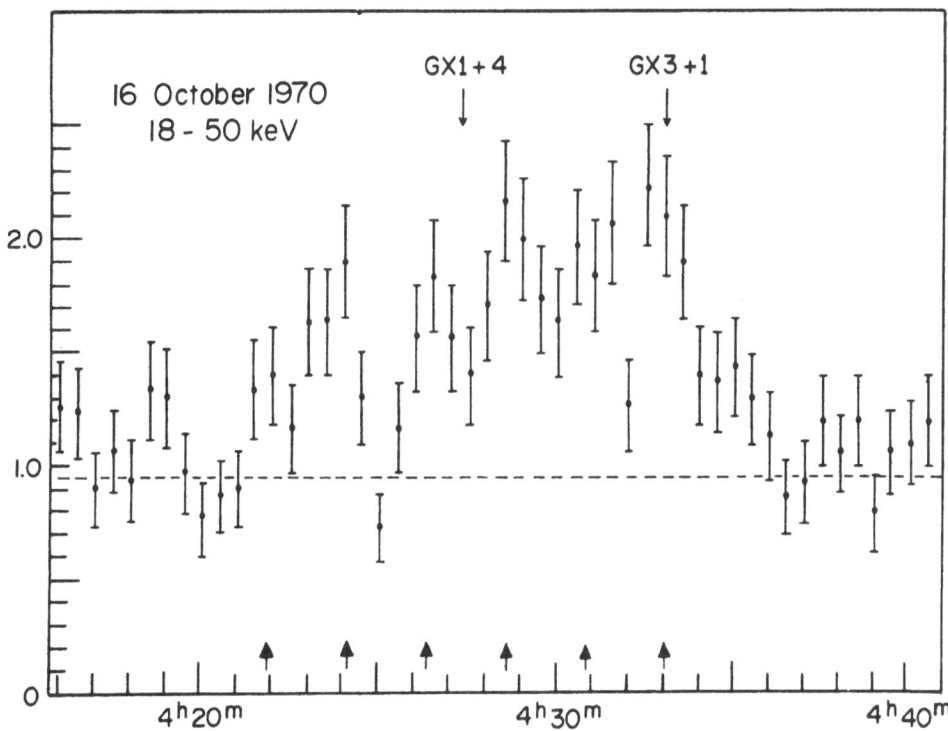

Figure 4   Discovery of the first slowly rotating neutron star in GX 1+4. A
series of X-ray flux changes were observed during balloon observations from
Australia in October 1970 . The data suggest changes with a period of about 2.3
minutes (see arrows pointing upward). Satellite observations made several years
later confirmed this discovery .   The rotation period of the neutron star is
decreasing (the neutron star spins up) and was about 1.8 minutes in 1982 .
(Courtesy of the author , G. Ricker , and J. McClintock , Astrophysical Journal
Letters , 1 October 1971 .)

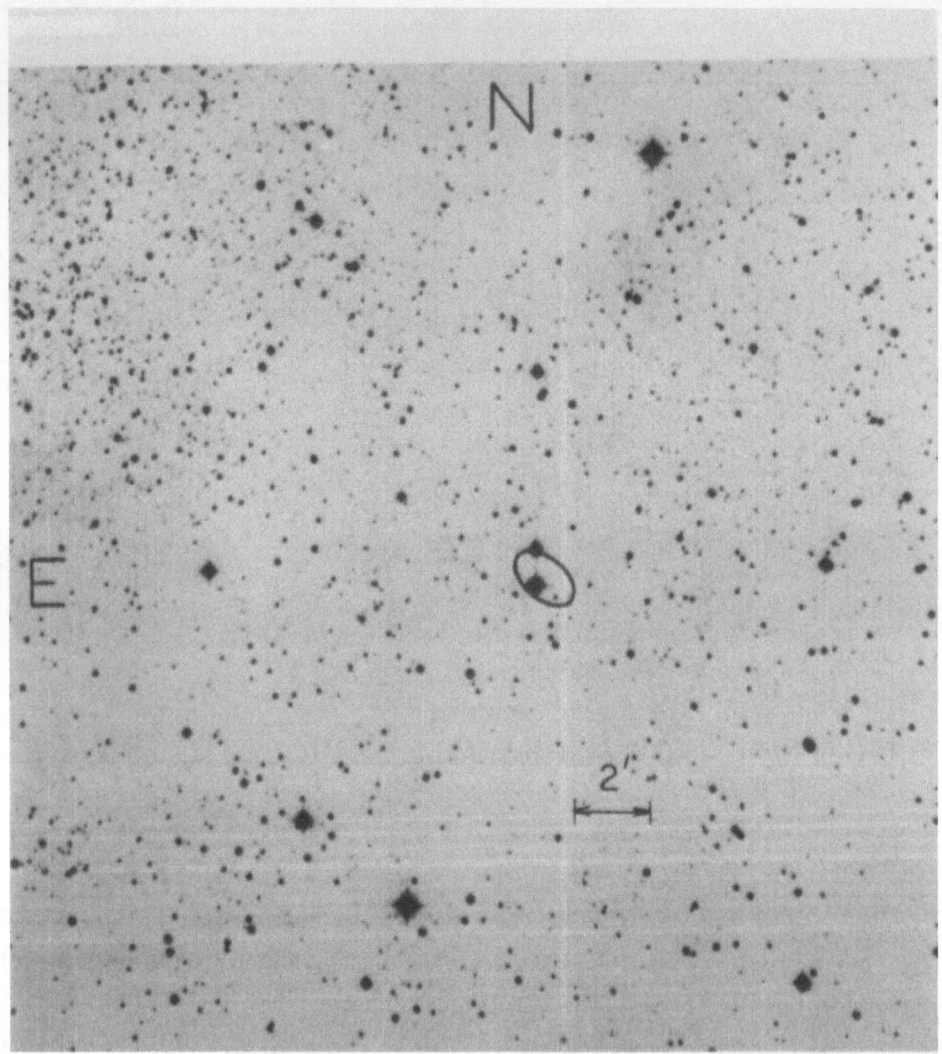

Figure 5   The binary star Cyg X-1 is the brightest star near the "bottom" of the ellipse . The light comes from a "normal" nuclear burning star . The compact X-ray object (invisible on this picture) is probably a black hole ; the orbital period of the system is 5.6 days .   (Courtesy of S. Rappaport , W. Zaumen , and R. Doxsey , Astrophysical Journal Letters , 15 August 1971 .)

We know that we are often dealing with neutron stars in these X-ray objects because the X-ray emission is often pulsed , just like the pulsed radio emission from pulsars (neutron stars). Figure 2 shows data from a source (called Cen X-3) whose X-rays are pulsed with an interval of 4.8 seconds . There is no question about the interpretation of this data ; we are seeing here the signature of the rotation of a neutron star .

How do we know that there is a companion star like the one shown in Figure 1 ? When the stars go around each other , it is possible that the neutron star , as seen from earth , will hide behind the "normal" star . When that happens the X-rays will be absorbed by that star and will not be received on earth ; we call that an **X-ray eclipse.** Figure 3 shows a recording of such X-ray eclipses as observed from Cen X-3 . Note the absence of X-rays every 2.1 days .

Neutron stars do not always rotate around in only a few seconds . In the same year (1971) when the pulsations of Cen X-3 were discovered by scientists from Harvard using data from the U.S. satellite **UHURU** , scientists from MIT discovered X-ray pulsations from a source they called GX 1+4 (the MIT scientists used data from balloon-borne X-ray detectors) . The intervals between the X-ray pulses were 2.3 minutes (see Figure 4) ; the neutron star in GX 1+4 rotated about its axis in about 2.3 minutes (it is speeding up rapidly , in 1982 it made one rotation in about 1.8 min) .

There is one case of particular interest where there are good reasons to believe that we are dealing not with a neutron star but with a **black hole .** This is the famous case of the source called Cyg X-1 , located in the constellation Cygnus (The Swan). Figure 5 shows the star (the brightest) inside the ellipse . This bright star is the donor that supplies the material to its compact companion star which is probably a black hole and not a neutron star . I do not want to go into the details here why that is believed .

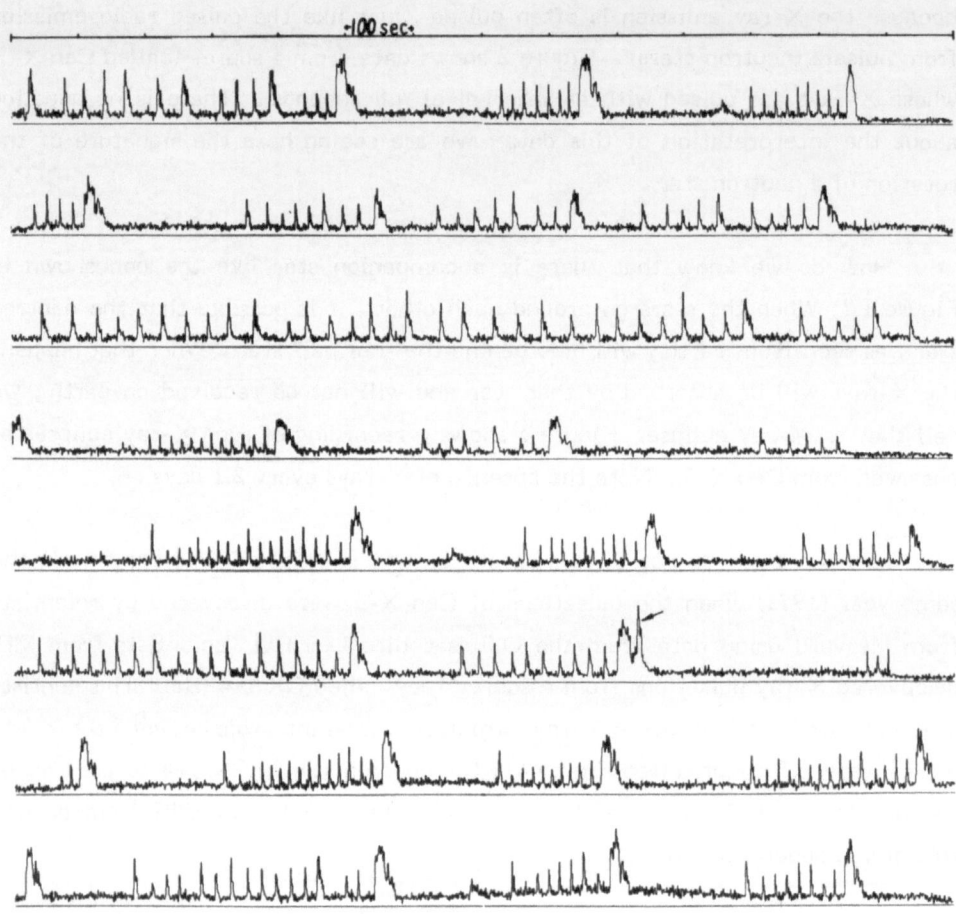

Figure 6   Discovery of X-ray bursts from the Rapid Burster (MXB 1730-335).
These are 24-minute stretches of data from eight different satellite orbits of the
SAS-3 observatory on 2-3 March 1976 when this unique object was discovered.
The bursts are the result of instabilities in the gas flow from a "normal" star onto
a neutron star (see Figure 1).   A burst active period lasts only a few weeks and
they come with approximately six-month intervals for reasons yet unknown.
(Courtesy of the author, Annals of the New York Academy of Science, 1977,
Volume 310, page 210 - Proceedings of the Eighth Texas Symposium on
Relativistic Astrophysics held in Boston, Massachusetts in December 1976.)

3. Gas-Flow Instabilities - X-ray Bursts - Optical Bursts - Nuclear Explosions

The flow of matter from the donor star to the neutron star (or black hole in perhaps a few cases) is not always steady but can sometimes be very erratic. There is an object known in our galaxy where we see this erratic, spasmodic accretion flow of material onto a neutron star in a very dramatic way. This object is called the **"Rapid Burster"**; it was discovered in 1976 with the U.S. SAS-3 observatory (see Figure 6). It seems that the gas falling towards the neutron star is first held up (probably by magnetic fields). Then, all of a sudden, a blob of matter "breaks through" and when it reaches the surface of the neutron star, a blast of X-rays is produced. This process repeats itself. We are dealing here with a very mysterious object indeed; the accretion instability which causes this erratic behavior is not understood. The Rapid Burster only produces these X-ray bursts (like machine gun fire) for a few weeks, then it stops and six months later it starts up again. That too is not understood.

The objects that I have discussed so far are called **"X-ray binaries"**. The X-ray energy is generated because the gas falls in a gravitational field; we call this release of gravitational potential energy (pardon me for the jargon). It is the same energy that is released when I drop a piece of chalk on the floor; it will fall, increase in speed, and break (and heat up) at impact.

This is not yet the end of the X-ray story for these bizarre binaries. Because of the gas flow (mass transfer) the gas (largely hydrogen) is accumulating on the surface of the neutron star; its temperature and density on the surface of the neutron star are very high. Thus the hydrogen can "burn" (nuclear fusion) to helium and the helium can "burn" to carbon. Under certain conditions, the helium fusion can become unstable; this is called a "runaway" or you could call it a nuclear bomb. The helium ignites and the helium layer on the neutron star (about a meter thick) fuses in a fraction of a second and releases a large amount of nuclear energy. It drives the temperature of the surface of the neutron star up to thirty million degrees and produces a burst of X-rays. The accumulation of matter on the neutron star continues and another burst can follow (in a few hours

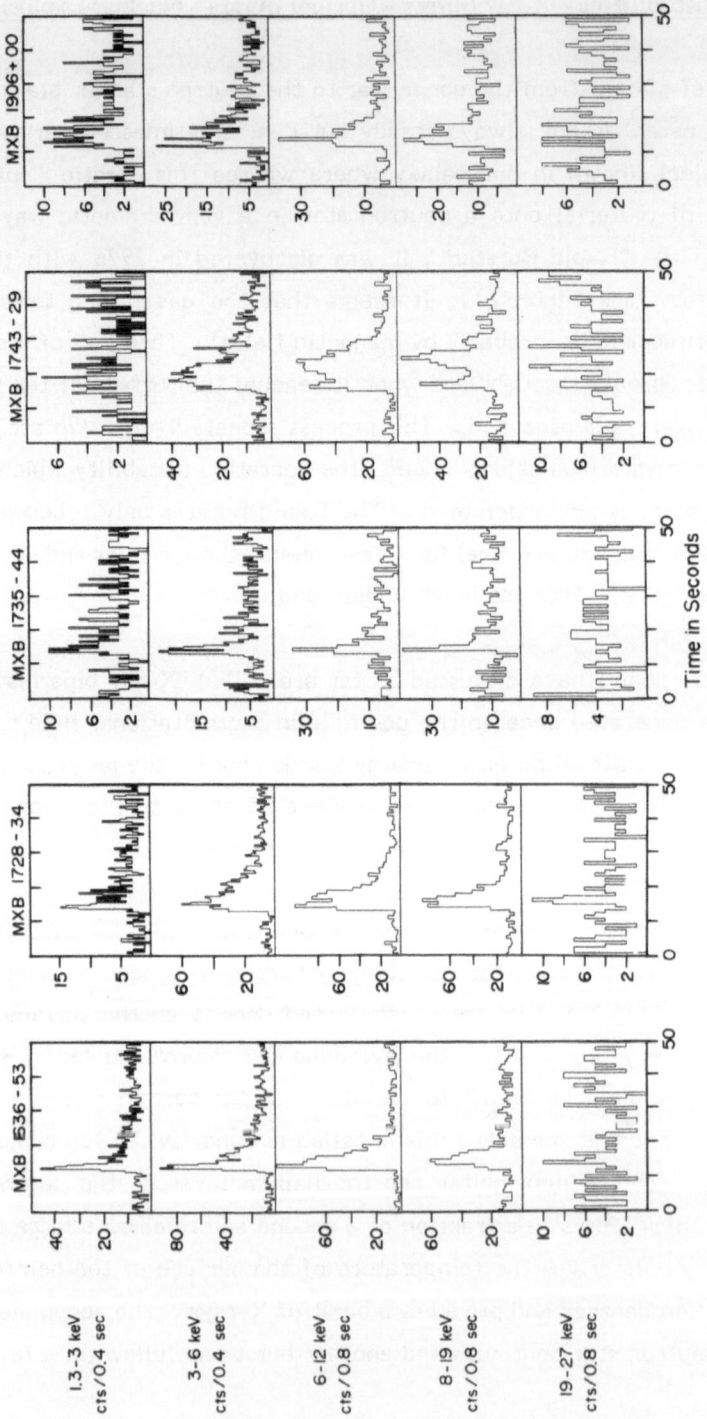

Figure 7

Figure 7   Time profiles of X-ray bursts from five different sources . (MXB stands for MIT X-ray Burst Source) . The X-ray data taken with the SAS-3 Observatory are shown for five energy bands as indicated on the left . The bursts are due to thermonuclear flashes (nuclear explosions) on the surface of a neutron star . The nuclear fuel is supplied by a nearby companion star (see Figure 1) . These bursts are of a very different nature than those shown in Figure 6 . (Courtesy of the author and P. Joss , Nature , 1977 , Volume 270 , page 211 .)

---

or a day or longer) . Thus , occasionally , in addition to the already discussed persistent X-ray emission a gigantic **X-ray burst** can occur . Figure 7 shows some X-ray bursts which are the result of nuclear bombs which explode on the surface of neutron stars . Note the sudden increase of the X-ray signal ; it can easily be ten times stronger during burst maximum than before the burst . These bursts (thermonuclear energy) should not be confused with those in Figure 6 which are the result of instabilities of matter falling onto the neutron star (gravitational potential energy) .

The burst of X-rays can produce (as a "byproduct") a blast in optical light . Some of the X-rays produced in the bursts are absorbed by the nearby companion star (and/or by the accretion disk around the neutron star) . This drives up the temperature of the companion star (and/or the disk) and increases the production of visible light . We have searched for years to see whether we could detect optical flashes from the ground (with optical telescopes) simultaneously with the X-ray bursts (detected with X-ray observatories in Earth orbit) . The first success came in 1978 . To date (summer 1983) there are several dozen optical bursts that have been seen simultaneously with X-ray bursts . These optical bursts can be quite big ; the optical star can become four times brighter in one second . In Figure 8 I show data on one of these optical bursts observed simultaneously with an X-ray burst .

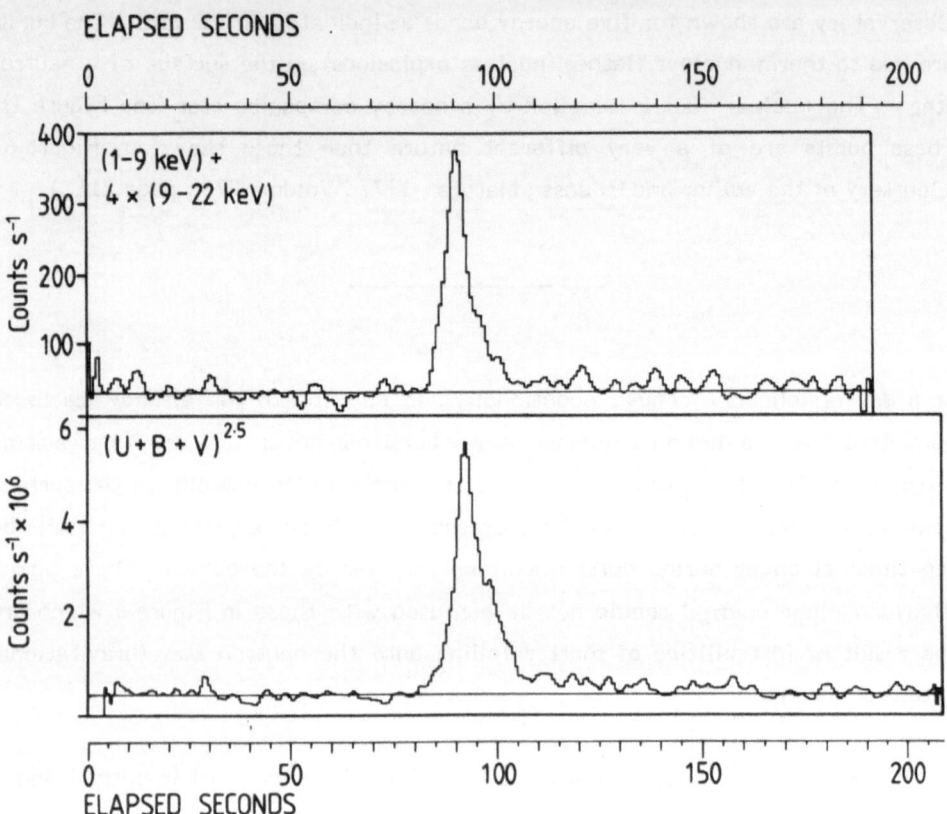

Figure 8   An X-ray burst (upper panel) from MXB 1636-53 as detected on 18 June 1980 by the Japanese X-ray observatory Hakucho . This burst was accompanied by an optical burst (lower panel) as observed with the 3.6m telescope of the European Southern Observatory at La Silla , Chile . The optical burst probably comes from an accretion disk around the neutron star (see Figure 1) . It is the result of a sudden heating caused by the X-ray burst (some X-rays from the burst are absorbed in the disk and heat it up) . (Courtesy A. Lawrence , L. Cominsky , C. Engelke , G. Jernigan , the author , M. Matsuoka , K. Mitsuda , M. Oda , T. Ohashi , H. Pedersen , and J. v. Paradijs , Astrophysical Journal , Volume 271 , 1983 .)

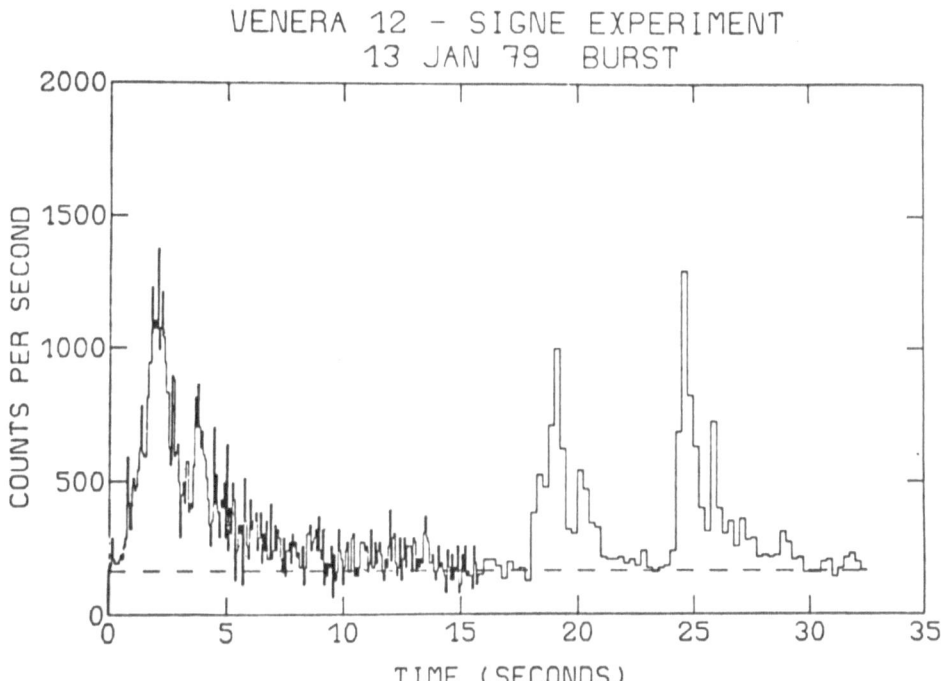

Figure 9   Gamma-ray burst detected on 13 January 1979 with the Russian Venera 12 satellite . The burst lasted about 30 seconds . Notice the complex time structure (e.g. , the three peaks) which is unlike that observed in X-ray bursts (see Figures 6 , 7 , and 8) . The nature of these bursts is still uncertain though it seems likely that they are caused by neutron stars . (Courtesy E. Mazets , S. Golenetskii , V. Ilyinskii , V. Panov , R. Aptekar , Yu. Guryan , I. Solokov , Z. Sokolova , and T. Kharitonova , Gamma-Ray Burst Catalogue , 1981 , Astrophysics and Space Science , Volume 80 , page 3 .)

4 . Gamma-Ray Bursts - Optical Bursts

There is another cosmic burst phenomenon which is much less understood . In the seventies scientists from Los Alamos using U.S. satellites of the **Vela** series discovered **gamma-ray bursts** . Figure 9 shows data on a gamma-ray burst which lasted 30 seconds and which took place on 13 January 1979 . The time structure is very complex (much more than observed in X-ray bursts) . The rise time of gamma-ray bursts can be a few milliseconds , thus much shorter than in the case of X-ray bursts which have typical rise times of one second . The gamma-ray bursts do not repeat on a time scale of hours or days as X-ray bursts do . The frequency of strong gamma-ray bursts from the gamma-ray burst sources is not well known ; it is , in general , less than one burst per few years . The gamma-ray burst sources are normally undetectable as gamma-ray sources ; (unlike X-ray burst sources which are also strong X-ray emitters between the bursts) . Largely due to the pioneering work of the Leningrad group in the Soviet Union we believe that gamma-ray burst sources are neutron stars . It is almost certain that the gamma-ray burst sources are in our own galaxy as the X-ray burst sources are .

It remains to be seen what causes these gamma-ray bursts . I would not want to speculate on their thermonuclear origin as some people have , but it seems safe to bet that the gamma-ray burst sources are very close to the earth , perhaps on the average no more than a few hundred light years away (the X-ray burst sources are much further away ; they are on the average at a distance of about 30,000 light years) .

On 19 November 1978 there was an unusually bright gamma-ray burst and through a network of many gamma-ray burst monitors in space , it was possible to pinpoint in the sky the location of that source to about 5 arc minutes square . An optical astronomer would tell you that a region of 5 arc minutes square , (which is only a very small fraction of the angular size of the moon) , is still very large and if you'd ask him/her to make observations to see whether there is something special in that region , he/she might laugh . If you look at that particular region in the sky with powerful telescopes , you would see hundreds of objects and wouldn't know which one to "pick" . It would indeed be very difficult to decide which one of these hundreds is the source of the gamma-ray burst .

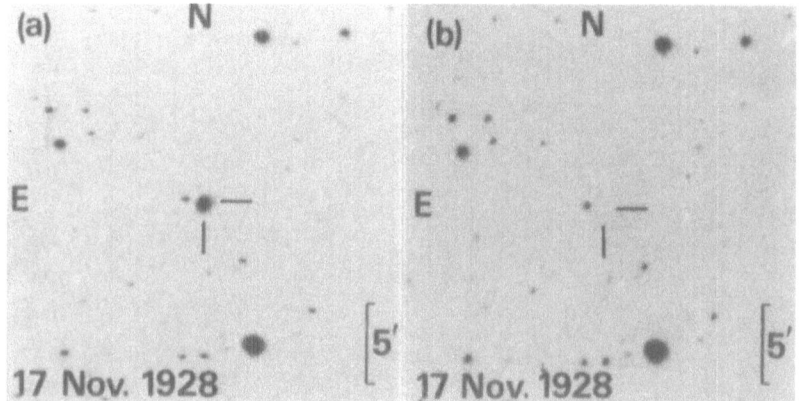

Figure 10   Discovery of an optical flash on archival plates . The figure shows two photographic plates taken in South Africa on 17 November 1928 . The plates were taken 45 minutes apart . The marks indicate the location of a bright object (left) , absent on the plate shown in the right panel . A gamma-ray burst was seen fifty years later from this region of the sky on 19 November 1978 , and it seems very likely that the optical object is associated with it (i.e. , that a gamma-ray burst also occurred on 17 November 1928) . (Courtesy B. Schaefer , Nature , 1981 , Volume 294 , page 722 .)

One of our graduate students at MIT , Brad Schaefer , at his own initiative , went to the Harvard archival plates and found about 5,000 plates that covered the small area in the sky (where the 19 November 1978 burst took place) taken since the end of the 19th century . He looked very carefully on all 5,000 plates and he found that on 17 November 1928 that region of the sky was photographed four times in a row (45-minute exposures each) , from South Africa . On one of those four plates he found a very bright "star" that is not visible on any of the other plates (see Figure 10) . It seems very likely that this was an optical flash that accompanied a gamma-ray burst . Thus if there had been a sufficiently sensitive gamma-ray observatory in space on November 17, 1928, it might have been observed .

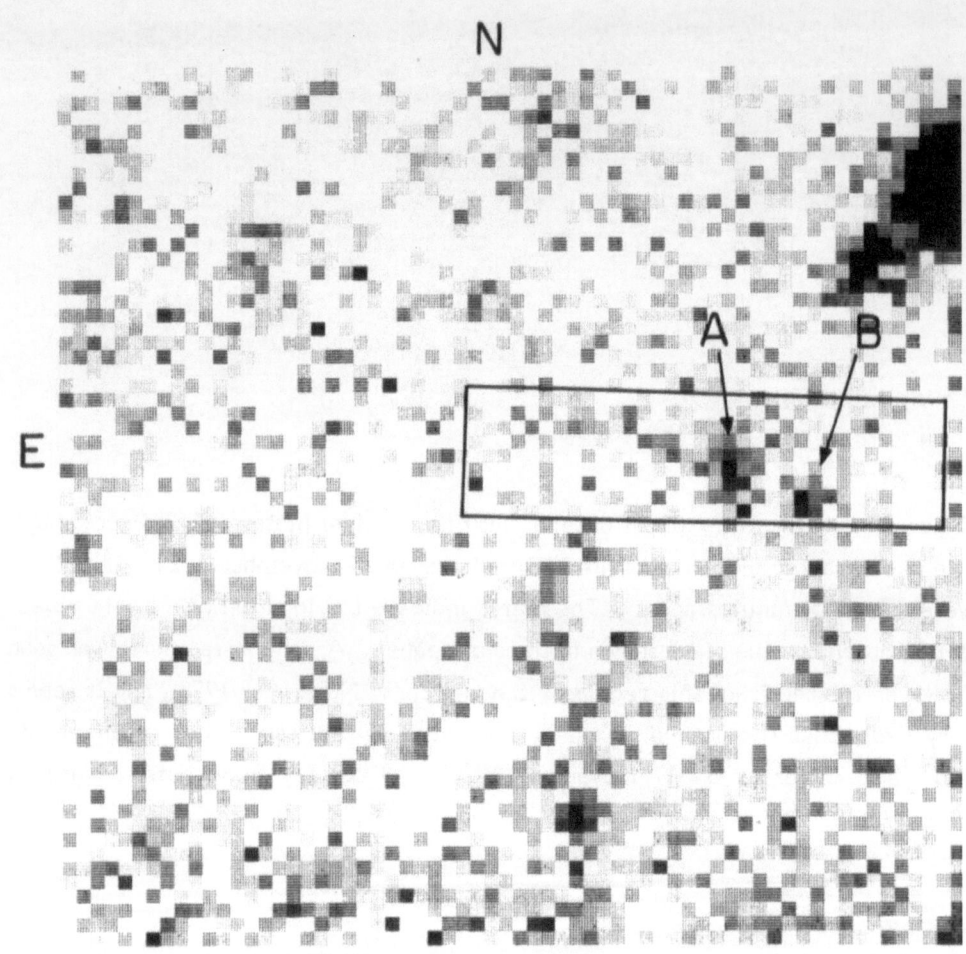

Figure 11   Close-ups of a small part of the region shown in Figure 10 made with the 1.5 meter Danish telescope of the European Southern Observatory in La Silla , Chile .

(a)   A combined 7.75 hour exposure taken in July 1981 and July 1982 . Two new objects were found , they are marked A and B . The optical flash as shown in Figure 10 occurred in the rectangular "box" as indicated here (with dimensions 4 arc sec by 16 arc sec) .

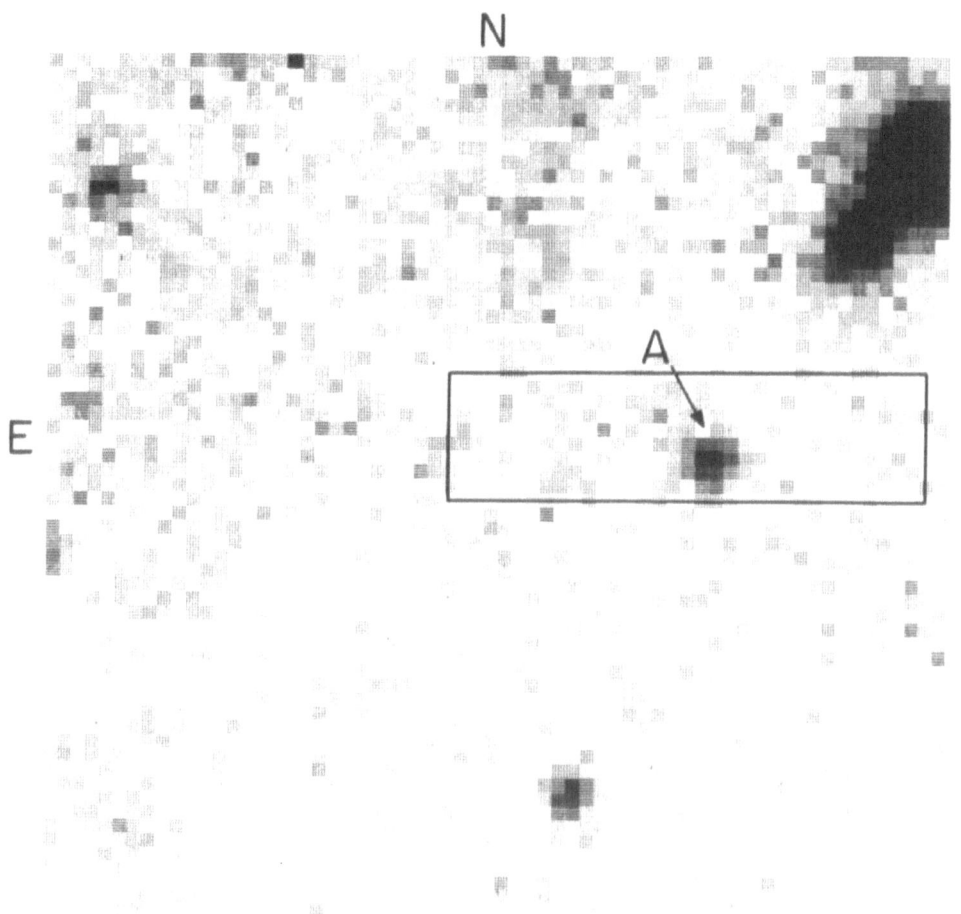

(b)   A combined 21 hour exposure taken in September to December 1982 . Source B is not visible anymore . Either A or B (or neither one) may be the source that caused the optical flash on 17 November 1928 , shown in Figure 10 , as well as the gamma-ray burst on 19 November 1978 . (Courtesy H. Pedersen , C. Motch , M. Tarenghi , J. Danziger , G. Pizzichini , and the author , Astrophysical Journal Letters , summer 1983 .)

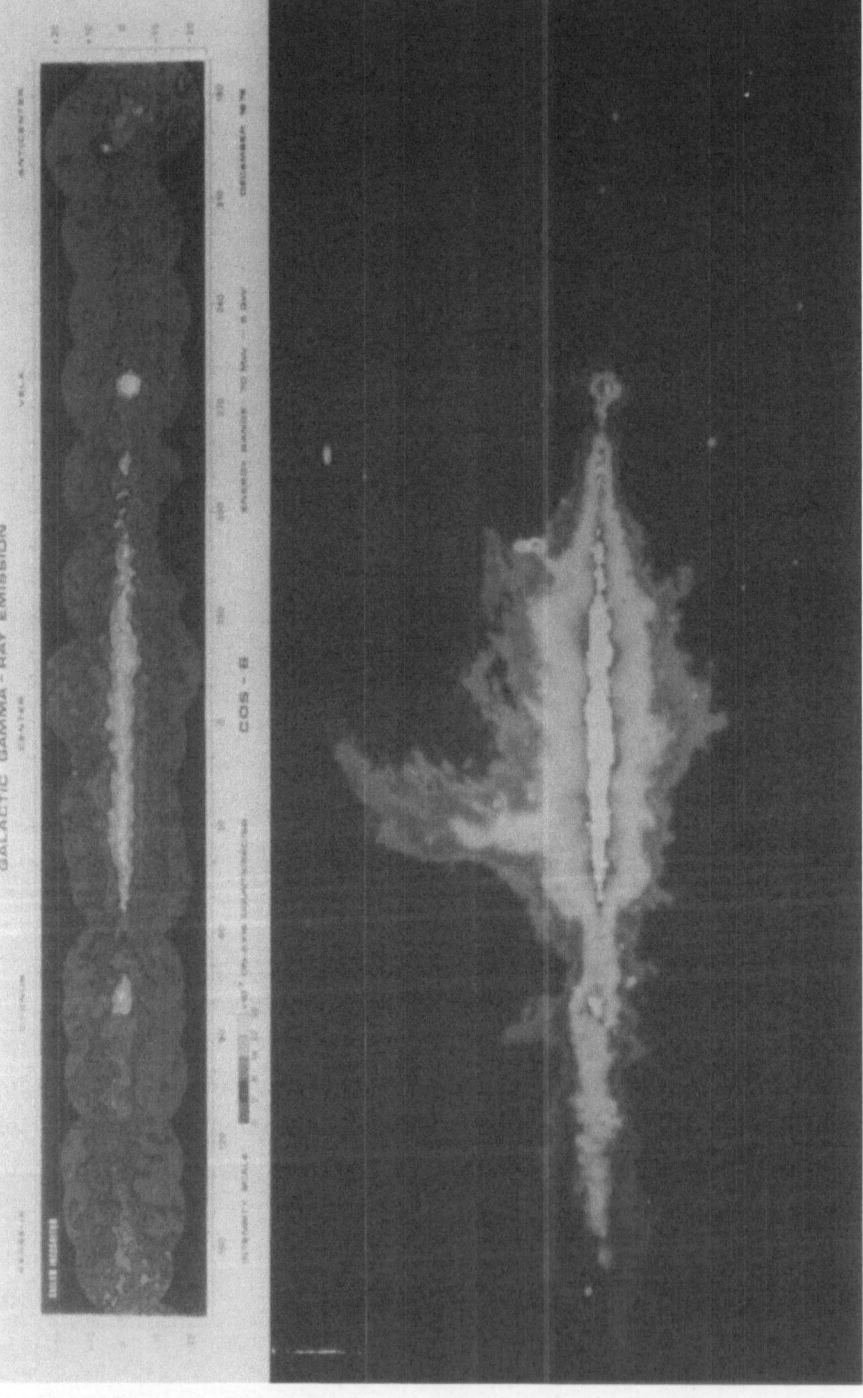

Figure 12

As recently as about one week ago (early August 1982) astronomers of the European Southern Observatory looked closely at the location of this "star" with very sensitive optical equipment . In the area , where the star was seen in 1928 , there are two very faint objects (see Figure 11) and it is quite possible that one of these is the gamma-ray burst source . I expect that in this decade we will learn much more about these enigmatic gamma-ray burst sources .

5 . Very High-Energy Gamma-Rays - Supernova Explosions

Cosmic gamma-rays were already discovered before the gamma-ray bursts in the early seventies through the pioneering work with the American OSO-3 satellite which was followed by another U.S. satellite called SAS-2 and later by the COS-B satellite of the European Space Agency . Some of the COS-B results are shown in Figure 12 . The gamma-ray sky is very bright in certain regions . If we look at a 21-cm radio emission (due to atomic hydrogen) map of the same scale we often note very good agreement between the radio emission and the

----

Figure 12   Two maps of the galactic equator region .

Top : A very high-energy gamma-ray (70 MeV–5 GeV) map as derived from data of the COS-B gamma-ray observatory . The lighter the color , the stronger the emission . The spatial scales (in degrees) are indicated .

Bottom : A 21-cm radio map of the same region of the sky (same scale) . The lighter the color , the stronger the emission . Notice the similarities between the two maps (see also text) . (Gamma-ray map courtesy of H.A. Mayer-Hasselwander ,   K. Bennett ,   G.F. Bignami ,   R. Buccheri , P.A. Caraveo ,   W. Hermsen ,   G. Kanbach ,   F. Lebrun ,   G.G. Lichti , J.L. Mansou , J.A. Paul , K. Pinkau , B. Sacco , L. Scarsi , B.N. Swanenburg , and R.D. Willis , 1982 , Astronomy and Astrophysics , Volume 105 , page 164 .   Radio map courtesy of H. Bloemen , Proceedings of Workshop on Gamma-rays held in Leiden , The Netherlands , ed. Burton and Israel) .

gamma-ray emission . It is therefore believed that some gamma rays come from the same hydrogen that is also responsible for the 21-cm radiation . In addition there are about 12 "point-like" sources that produce very high-energy gamma-rays , one of which I will mention shortly . Before I do that , I would like to say a bit more about neutron stars .

It was suggested in 1933 (one year after the neutron was discovered) that when stars die (i.e. , when their nuclear fuel is used up) the star's inner part (the core) can collapse while the outer part explodes and in the process of the core collapse a neutron star is formed . The explosion can be so powerful that during the early weeks after such an explosion the star can shine in optical light as a hundred billion stars do together . Since there are roughly a hundred billion stars in a galaxy , one such a **supernova** explosion would be about as bright as all other stars in its galaxy . As you can see in Figure 13 , the amount of light from the supernova in NGC 5253 is nearly equal to the light from the whole galaxy in which it is located .

The best known supernova explosion took place in the constellation of Taurus (the Bull) on 4 July 1054 (more than 900 years ago) . It was observed by Chinese astronomers but strangely enough , it was not recorded by the Europeans . The star was so bright that during the day it was visible for several weeks and for at least a year it was the brightest star in the sky . Hundreds of years later , in the 19th century , it was seen again (with the use of a telescope) as a "Crab-like" Nebula ; it can be seen in Figure 14 . The Nebula is about five arc minutes in diameter (i.e. , 1/6 the angular size of the moon) and it is at a distance of about

---

Figure 13   Supernova explosion in the galaxy NGC 5253 . The four pictures were made over a period of about one year . The brightness of the supernova gradually decreased : 16 May 1972 (top left) ; 24 April 1973 (bottom right) . During a few weeks the star was nearly as bright as all stars in the galaxy . These pictures were made by Charles T. Kowal with the 48-inch Schmidt telescope on Mount Palomar .

Figure 14   The Crab Nebula is the remnant of a supernova explosion that occurred on 4 July 1054 when the sudden appearance of a "guest star" was recorded by Chinese astronomers . During several weeks the star was even visible during the day . The gas filaments , visible in the photograph , are streaming out with velocities of a few thousand kilometers per second . The star that exploded is the most southern one of the two near the center of the Nebula ; it io a neutron star with a density about 10,000,000,000,000 times that of our Earth ; it rotates about its axis in about 33 milliseconds and is about 20 km across . The distance to the Crab Nebula is about 5,000 light years . The Nebula is about eight light years across . The photograph was taken using a red filter and the 200-inch telescope of Palomar   Mountain .   (Courtesy   of   The   Hale   Observatories ,   Pasadena , California .)

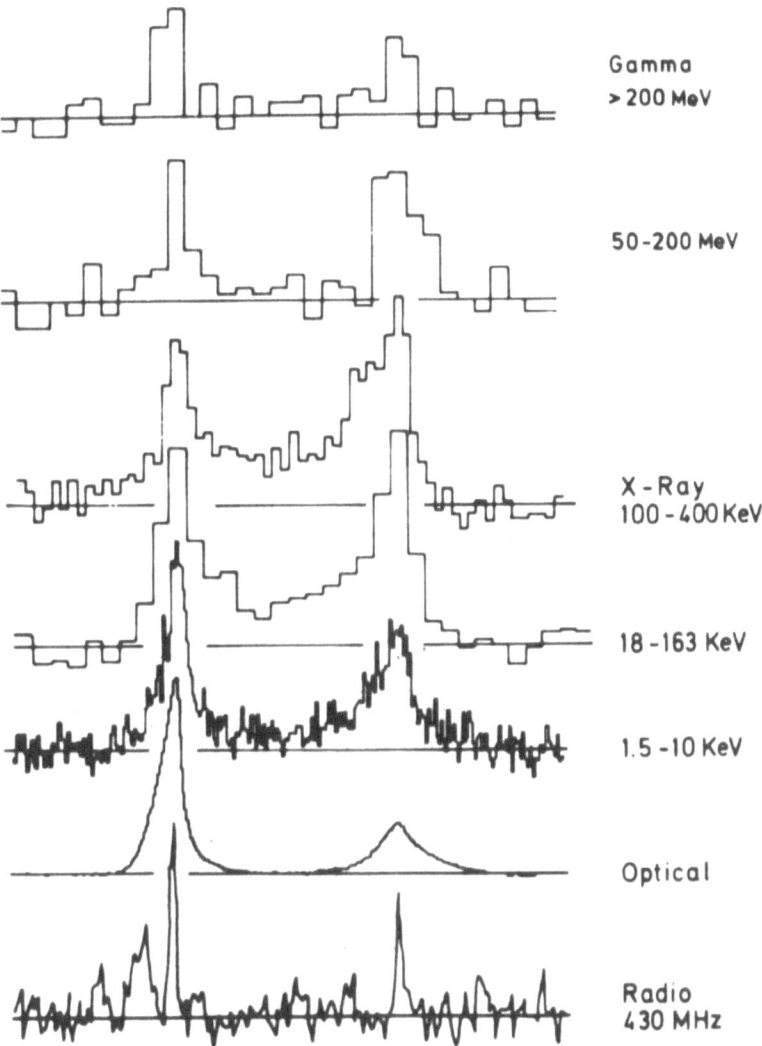

Figure 15   "Light-curves" of the pulsar in the Crab Nebula of electromagnetic radiation all the way from the radio emission (bottom) up to the high-energy gamma-rays (top).   The horizontal axis represents the period of precisely one rotation of the pulsar (about 33 milliseconds). (Courtesy of G. Bignami and W. Hermsen , Annual Review of Astronomy and Astrophysics , 1983 , Volume 21 .)

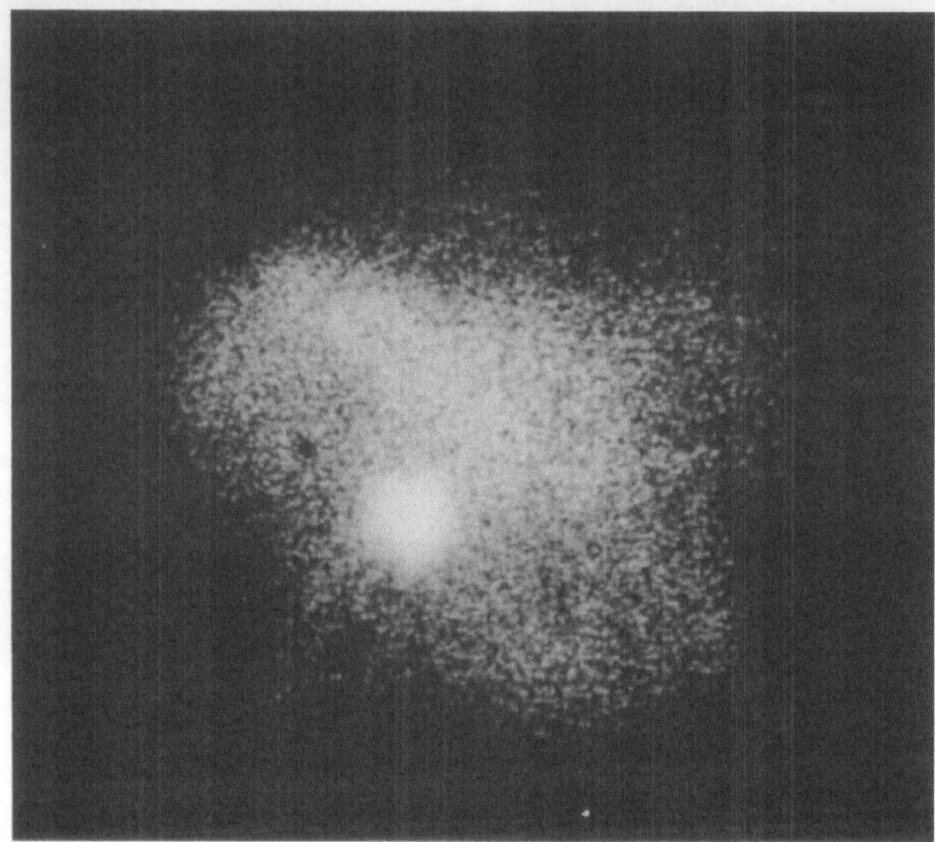

Figure 16   X-ray picture of the Crab Nebula taken with the Einstein Observatory . This picture looks very different from the one in Figure 14 which was taken in the optical . The pulsar stands out in this X-ray picture like a sore thumb ; it is inconspicuous in the optical . The filament structure , very obvious in the optical photograph , is completely absent in this X-ray picture . The scale (angular size per inch) of this picture is about three times that of Figure 14 . (Courtesy of F.D. Harnden , Jr. , Harvard/Smithsonian Center for Astrophysics ; article by R. Giacconi and H. Tananbaum , Science , August 1980 , Volume 209 , 'No. 4459 , page 865 .)

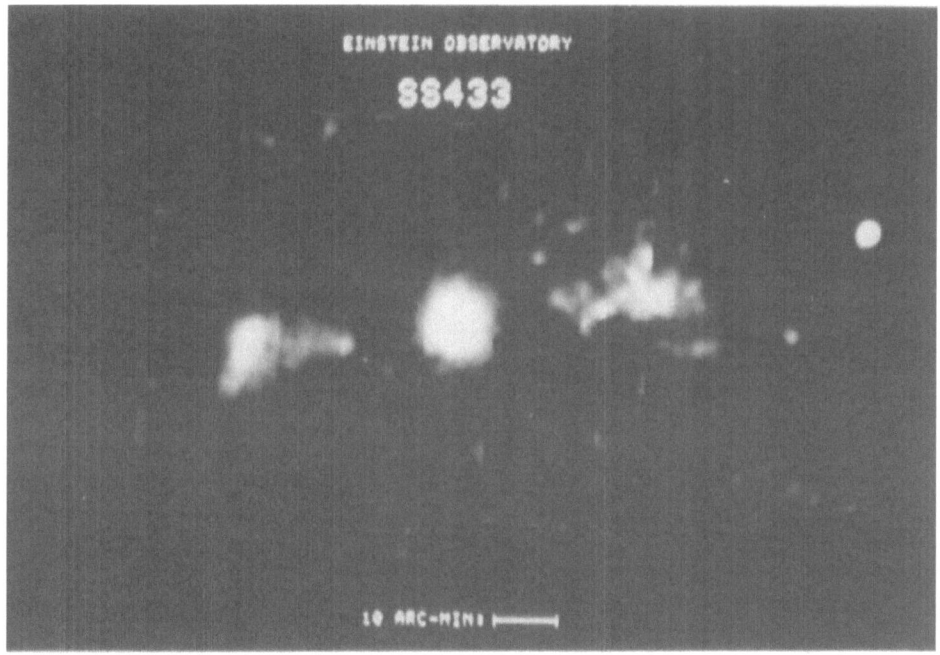

Figure 17   SS 433 in X-rays. This picture was taken with the Einstein Observatory. Two jets are visible, they have velocities of about 50,000 km per second. This may be a "pocket-size" prototype for the giant jet structures sometimes seen in active galaxies (see Figures 18-20).   (Courtesy of F.D. Seward, Harvard/Smithsonian Center for Astrophysics ; to appear in The Astrophysical Journal , 1983 .)

Figure 18   Optical photograph of the galaxy Centaurus A .   Notice the obscuration due to dust (dark bands).   This galaxy is at a distance of about 15 million light years . This is an active galaxy ; it produces jets (lobes) as shown in Figures 19 and 20 (Courtesy of European Southern Observatory).

5,000 light years . The gas filaments that you can see , are due to the hydrogen from the star that exploded . It is a neutron star (visible in the picture ; see caption for Figure 14) ; it rotates very fast about its axis ; about 30 times per second . The visible light , the X-rays and gamma-rays all show the signature of rotation of the neutron star (see Figure 15) . Figure 16 is an X-ray picture of the Crab Nebula taken with the U.S. Einstein Observatory . We can clearly see the bright pulsar as a "point-like" X-ray source in the X-ray emitting nebula .

6 . X-ray and Radio Jets

Before I take you farther out into the universe , I would like to dwell upon an object which has recently received a lot of interest ; it is called **SS 433** . It is a binary system : a neutron star and a "normal" companion star . Matter flowing onto the neutron star is producing X-rays , but it is doing something else : it produces two jets . Figure 17 shows these jets in X-rays . We know that the velocity of the jets is about 50,000 kilometers per second ! It is conceivable that SS 433 is a prototype (on a small scale) of the jets that we see farther out in our universe near very active galaxies .

We are now leaving our own galaxy and travelling a distance of about 15 million light years . On the way we meet galaxies like our own : very normal and very modest . We always like to think of ourselves as normal and modest , and when something is different we call it strange or abnormal . Some galaxies are "abnormal" indeed in the sense that near their centers they produce much more energy than our galaxy does near its central region ; we call them **active galaxies** . It is quite possible that massive black holes reside at the centers of these active galaxies . These black holes probably "feed" themselves by swallowing stars that are getting too close . In some cases the active galaxies show **jets** . There is such an object in the constellation Centaurus : radio astronomers call it Centaurus A . Figure 18 shows what it looks like photographed with an optical telescope ; it is an elliptical galaxy with a dust band , the diameter of the bright portion is about seven arc minutes . In this picture there is no obvious indication for it being extraordinary or very active and there are no jets (lobes) visible . If , however , we look with a radio telescope (Figure 19) we

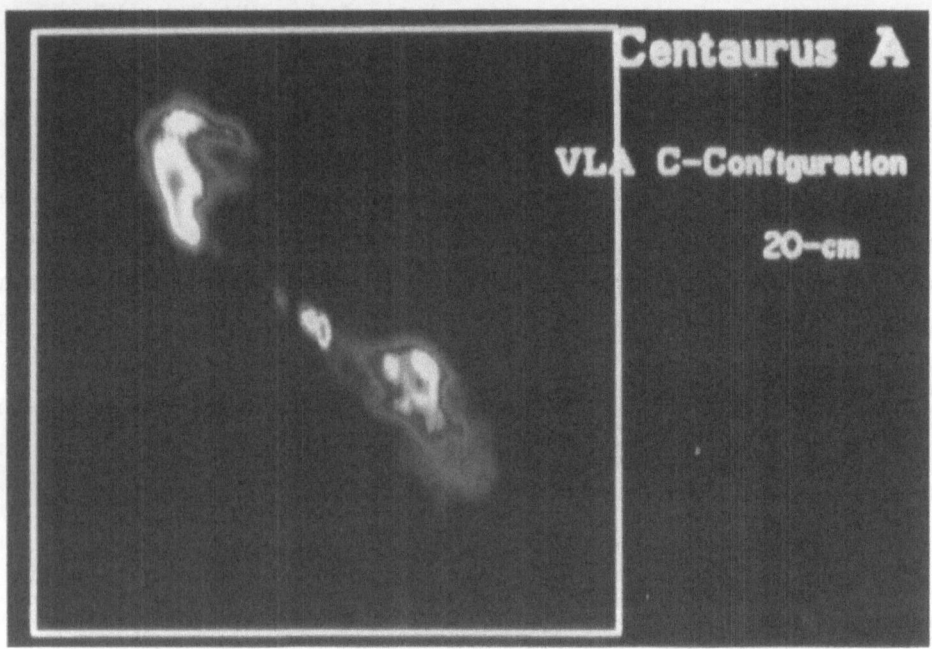

Figure 19   Centaurus A in 20-cm radio waves observed with the Very Large
Array in Socorro , New Mexico . Two jets (lobes) are visible . The extent of the
two combined as projected on the sky is about seven arc minutes . Thus they have
about the same size as the diameter of the galaxy shown in Figure 18 . Notice the
striking difference in appearance of this galaxy when observed in radio emission
compared to that in optical light (Figure 18) or in X-rays (Figure 20) . (Courtesy
of E. Schreier , J. Burns , and E. Feigelson , Astrophysical Journal , 1981 , Volume
251 , page 523 .)

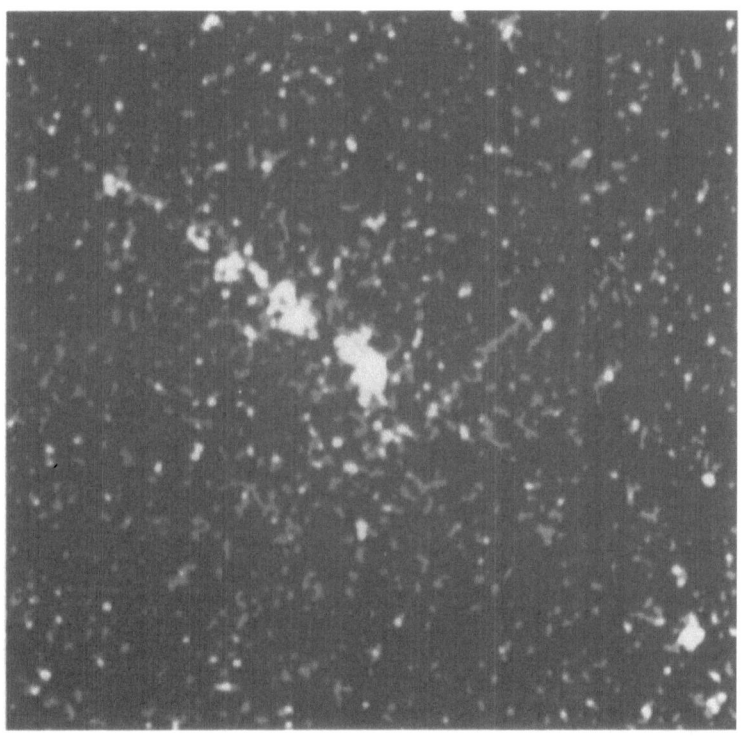

Figure 20   Centaurus A in X-rays . This picture was taken with the Einstein Observatory . Only one jet is visible . The scale of this photograph is comparable to that of Figure 19 .   (Courtesy of E. Feigelson , E. Schreier , J. Delvaille , R. Giacconi , J. Grindlay , and A. Lightman , Astrophysical Journal , 1981 , Volume 251 , page 31 .)

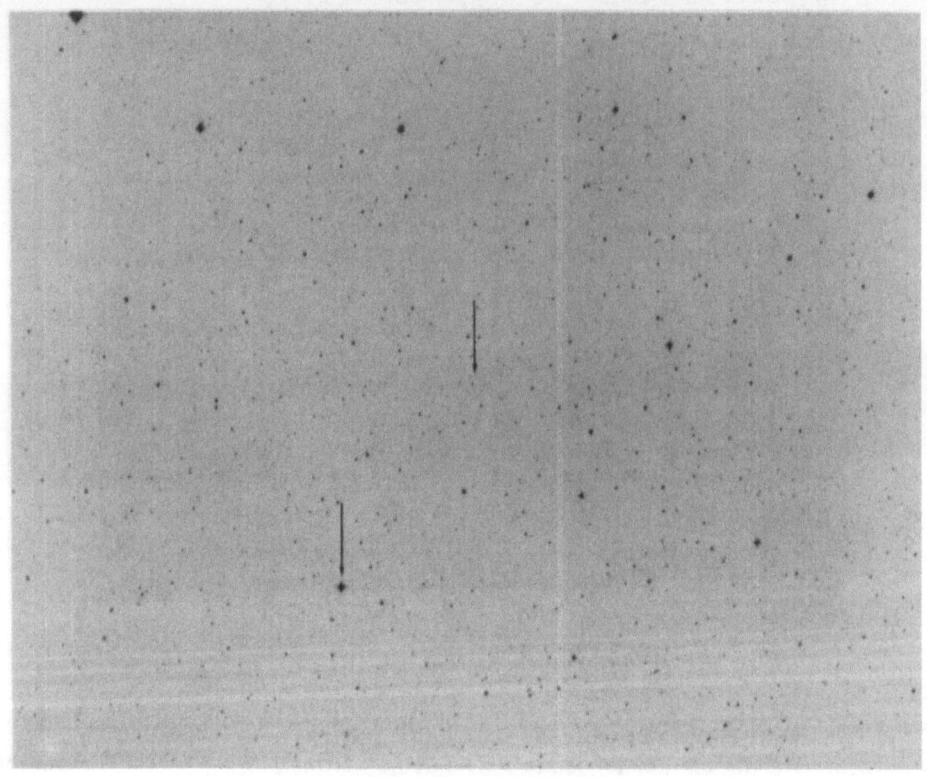

Figure 21   Optical photograph (same scale and orientation as Figure 22) marking the quasar 3C 47 and an X-ray emitting star at the tips of the two arrows . The star is approximately 5,000 times brighter than the quasar in visible light . (Courtesy of H. Tananbaum , National Geographic Society , and Palomar Observatory Sky Survey .)

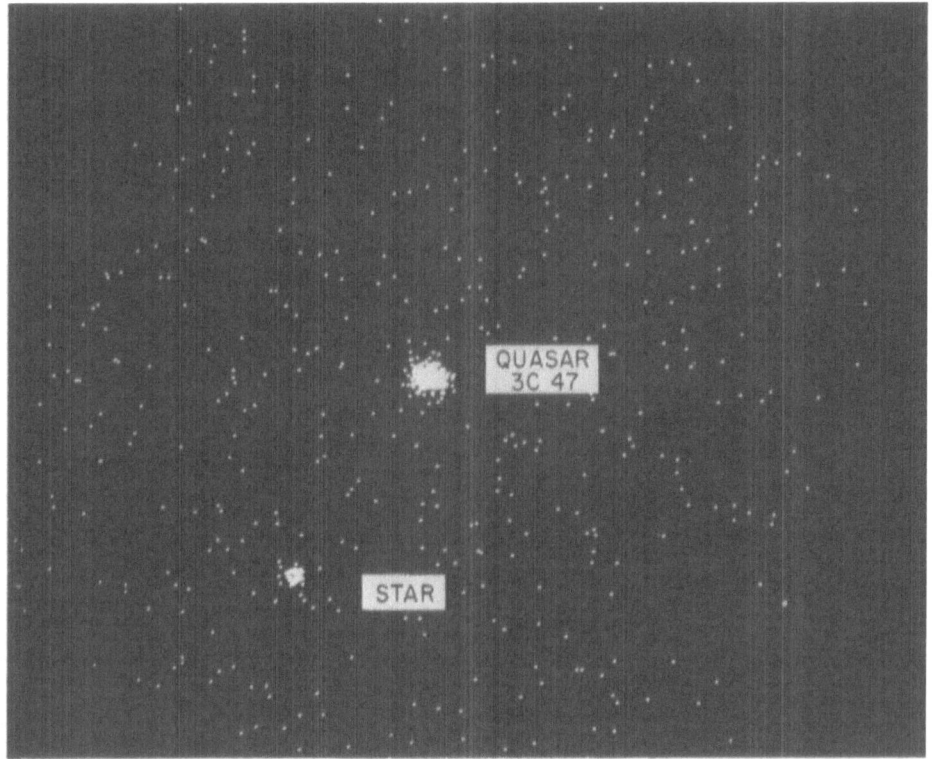

Figure 22   X-ray picture made with the Einstein Observatory (same scale and orientation as Figure 21). The quasar 3C 47 is the bright X-ray object near the center . A second object identified with a relatively bright star , marked on Figure 21 , is also seen . Notice how the quasar stands out here and how it compares with its "modest" optical appearance in Figure 21 . (Courtesy of H. Tananbaum , Harvard-Smithsonian Center for Astrophysics .)

see that two lobes are present, however, on a similar picture in X-rays (Figure 20), there is only one. The optical picture apparently gives only a small part of the story; the radio picture alone does not give the whole story either, nor does the X-ray picture. It is the combination of all three that may allow for an explanation of these jets (lobes).

No one has yet succeeded in composing a theory that can explain the jets (lobes) in a satisfying way; they are each at least 20,000 light years long (about one fifth the size of our entire Galaxy). The jet radiation is probably the result of synchrotron radiation (high-energy electrons spiralling in a magnetic field). One of the outstanding problems in astrophysics is how the electrons are accelerated (you have to keep accelerating them otherwise the jets could not be as long as they are).

7.   Quasars (How to Find Them ?)

If we go further out into the universe we reach **quasars** which are also active galaxies. They are very powerful indeed: their centers radiate a billion times more energy than a comparable volume in our own galactic center region. Most of the quasars are quite far away from us. Because of the expansion of the universe, their velocities, in many cases, approach the speed of light. It is quite a task for optical astronomers to find these quasars; they are in general blue objects and they are recognized by their high velocities away from us (Doppler shift).

I would like to show how easy it sometimes is in X-ray observations to find quasars. Figure 21 is an optical picture of the sky with hundreds of objects. You wouldn't know which of these objects is a quasar unless you studied them to see which one meets the description of a quasar. If, however, we look at the same area in the sky in X-rays (Figure 22) you see that the quasar stands out like a sore thumb; it's the brightest one of them all, yet that object is the farthest of them all. If one wants to find more quasars, perhaps the easiest is to look first with X-ray telescopes and then with optical telescopes to see whether the object has the typical quasar characteristics.

8 . Summary

I have taken you on a journey which started with the sun , we moved into our galaxy and far beyond it . The space age has led us to objects which we could not dream of before . It has brought us the neutron star X-ray binary systems ; it has given us the X-ray and gamma-ray bursters (the latter we do not understand yet) and X-ray and gamma-ray active galaxies and quasars . I could make the list much longer .

X-ray and gamma-ray astronomy are very rich fields . They were born in the U.S. in the sixties and have prospered ever since . Many countries have contributed (notably England , the Soviet Union , Holland , Japan , and the European Caravan consortium under the direction of ESA) ; others are undoubtedly to follow . It hurts that there are presently no major U.S. X-ray or gamma-ray observatories in orbit ; Japan is presently (August 1982) the only country to have an X-ray observatory , called Hakucho . In 1983 Japan and Europe will each launch an X-ray observatory (called ASTRO-B and EXOSAT); and Germany and Japan will launch X-ray observatories (ROSAT and ASTRO-C) later in this decade ; the U.S. plans to launch a gamma-ray observatory (GRO) about that same time . New ambitious U.S. missions are planned for the late eighties and early nineties (the most ambitious of all is called AXAF) . They will certainly bring new surprises and our knowledge will penetrate further into the bizarre environments of cosmic laboratories where we encounter extreme conditions (e.g. , very strong magnetic fields , exceedingly high temperatures , and stellar objects with extraordinarily high densities) which could never be made or simulated on Earth . We will learn more about the "world" around us and we are bound to learn new physics ; that is what science is all about .

CHAPTER V

SPACE SCIENCE AND COSMOLOGY

by M.S. Longair

Royal Observatory , Blackford Hill
Edinburgh , Scotland , UK

Summary

Space Science has opened up many new wavebands for cosmological research during recent years . The new data , apart from being of great importance in the interpretation of ground based results , also provide new challenges for the cosmologist . At the same time , they indicate new ways of tackling classical cosmological problems . In this non-technical survey , these recent advances will be discussed in the context of our current understanding of cosmology and we will look to the advances which can be expected to be made by space astronomy in the next 20 years .

1 . Introduction

By **cosmology** , I mean the study of the properties of the Universe on the largest scale . The aim of scientific cosmology is the <u>study of the origin of the large scale structure of the Universe</u> and how all its contents came into being . These are among the most profound questions of modern science .

R. M. West (ed.), Understanding the Universe, 129–225.

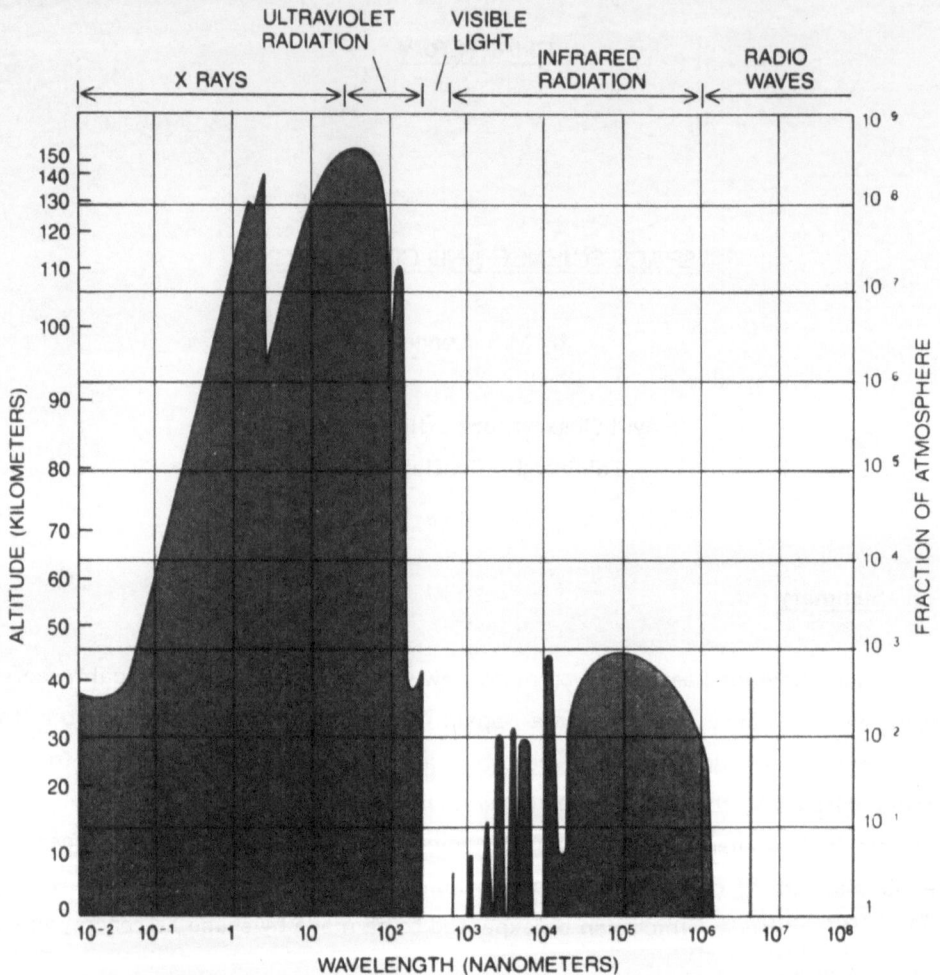

Figure 1   The transparency of the Earth's atmosphere to electromagnetic radiation i.e. radio , infrared , optical , ultraviolet , X and gamma-ray radiation . The shaded portions indicate altitudes at which radiation incident on the top of the earth's atmosphere is severely attenuated by atmospheric absorption (from J. Bahcall and L. Spitzer , 1982 , Scientific American , 247 , 40) .

Over the last twenty years , there has been a revolution in the study of cosmology . I would characterise it by stating that up till about 20 years ago , there were very few facts about which one could really be certain and there was much room for cosmological speculation . Nowadays , we have many more real facts about the Universe as a whole and a more secure framework within which to ask much more difficult physical questions about the origin of structure in the Universe .   To put it simply , the subject has changed from cosmological speculation to the astrophysics of the hot big-bang model of the Universe . This great advance has enabled astronomers and cosmologists to ask questions which it would not even have been possible to dream about 20 years ago and also to have realistic hopes of finding the answers .

Space science has provided some crucial pieces of information in the development of this new understanding and I will highlight them in this review. My own view is that many of the most important questions to which we need answers will be provided by space observations . Already we know the observing facilities which are needed to advance the subject . Telescopes which 20 years ago were no more than dreams can now be planned in detail with full consciousness of their fundamental scientific importance .

This survey will be in six sections . These will describe :

i)   the contents of the Universe ;

ii)   the case for the hot big-bang model of the Universe ;

iii)   the problems of galaxy formation ;

iv)   the astrophysical evolution in the Universe ;

v)   future space experiments of fundamental significance for cosmology ;

vi)   the fundamental problems of cosmology .

Before we begin this programme , let us look at the limitations of ground based astronomy . Figure 1 shows the altitude to which one must travel in order to obtain a clear view of the Universe at all wavelengths from the radio waveband to the hard X-ray and gamma-ray bands . Along the abscissa is plotted the wavelength of the radiation and the shaded areas indicate altitudes at which it is impossible to view the Universe because of absorption of radiation in the Earth's atmosphere . It can be seen that only in the radio waveband and in the rather narrow optical "window" is it possible to get a clear view of the universe from the surface of the Earth . The near infrared waveband has a number of narrow windows which are useful from high mountain sites . However , for all other wavebands , it is necessary to get above the Earth's atmosphere and this means observations from space . These wavebands , which can only be studied from space , are the submillimetre and far infrared wavebands , the ultraviolet , X and gamma-ray wavebands .

When we look in any single waveband , we get a highly selective view of the Universe . To give a simple example , in the optical waveband , the brightest objects we see are the stars and these are very nearby objects in astronomical terms . On the other hand , in the radio waveband , many of the brightest objects in the sky are very distant objects indeed . Some of them are among the most distant known objects in the Universe . The physical processes responsible for the optical emission of stars and the radio emission of the very distant radio sources are totally different and give us totally different information about the objects which make up our Universe .

We are only beginning to obtain a picture of what the Universe looks like in the wavebands which are the province of space astronomy . In the X-ray waveband , we now know that the brightest objects in the sky are binary X-ray sources (binary stellar systems in which one star is very compact) , exploding stars , active galaxies and quasars . We have already seen the very brightest objects in the gamma-ray waveband and most of these are probably related to the X-ray binaries . In the infrared waveband , we see many cool stars but we do not yet have a complete sky survey to know how common other cool objects are in the Universe . In the ultraviolet waveband , we do not yet have any detailed pictures but there must be many hot objects such as massive stars , white dwarfs and other hot sources which we have not yet thought of .

Until we have a detailed picture of the sky in all wavebands, we cannot claim to have a complete picture of the contents of the Universe. Fortunately, as we will show, astronomers and space scientists have defined the space observatories we need to achieve exactly these goals. As I have emphasised, each waveband has its distinctive character and provides complementary information about our Universe.

2. The Contents of the Universe

Perhaps the most frequent question I hear from non-scientists about astronomy and cosmology is "How can you imagine such vast distances?" The answer is really very simple. One never thinks in absolute terms but rather about one distance relative to another. Let me illustrate this by taking you from the size of the solar system to the edge of our observable Universe in six easy stages, each time going up in a scale by factors which are quite understandable in human terms. The relevant steps in this journey are shown in Table I.

We are all familiar with the distance between the Sun and the Earth. It is about 150 million km. Light travels at a speed of about 300,000 km per second and therefore light takes about 8 minutes to travel from the Sun to the Earth. It is often convenient to measure these large distances in terms of the time it takes light to travel that distance and so we can say that the Sun is at a distance of about 8 light minutes from the Earth.

Distances within the Solar System are now familiar to us since we have already sent space probes to Venus, the Moon and Mars and spacecraft have flown past the more distant planets, Mercury, the closest to the Sun, and the giant outer planets, Jupiter and Saturn. The distance from the Sun to Saturn is about ten times the distance from the Sun to Earth, i.e. a distance of about 80 light minutes. Because space probes such as Voyager I and II travel much slower than the velocity of light, the journey to Saturn from Earth took about 4 years.

TABLE I

THE CONTENTS OF THE UNIVERSE AND THEIR PHYSICAL DIMENSIONS

	Distance or Size	Relative Size
1. **Solar System** Sun to Saturn	$1,400,000,000 = 1.4 \times 10^9$ km	10 times the distance from Sun to Earth
2. **Nearest Stars** Sun to nearest stars	3×10^{13} km	20,000 times the distance from Sun to Saturn
3. **Our Galaxy**	3×10^{17} km	10,000 times the distance from Sun to nearest stars
4. **A cluster of galaxies**	about 3×10^{19} km	100 times the size of a galaxy such as our own
5. **Distance between rich giant clusters**	about 3×10^{20} km	10-20 times size of a cluster of galaxies
6. The **"Observable Universe"** at the present epoch of cosmic time	about 1.5×10^{23} km	about 50 times the distance between rich clusters

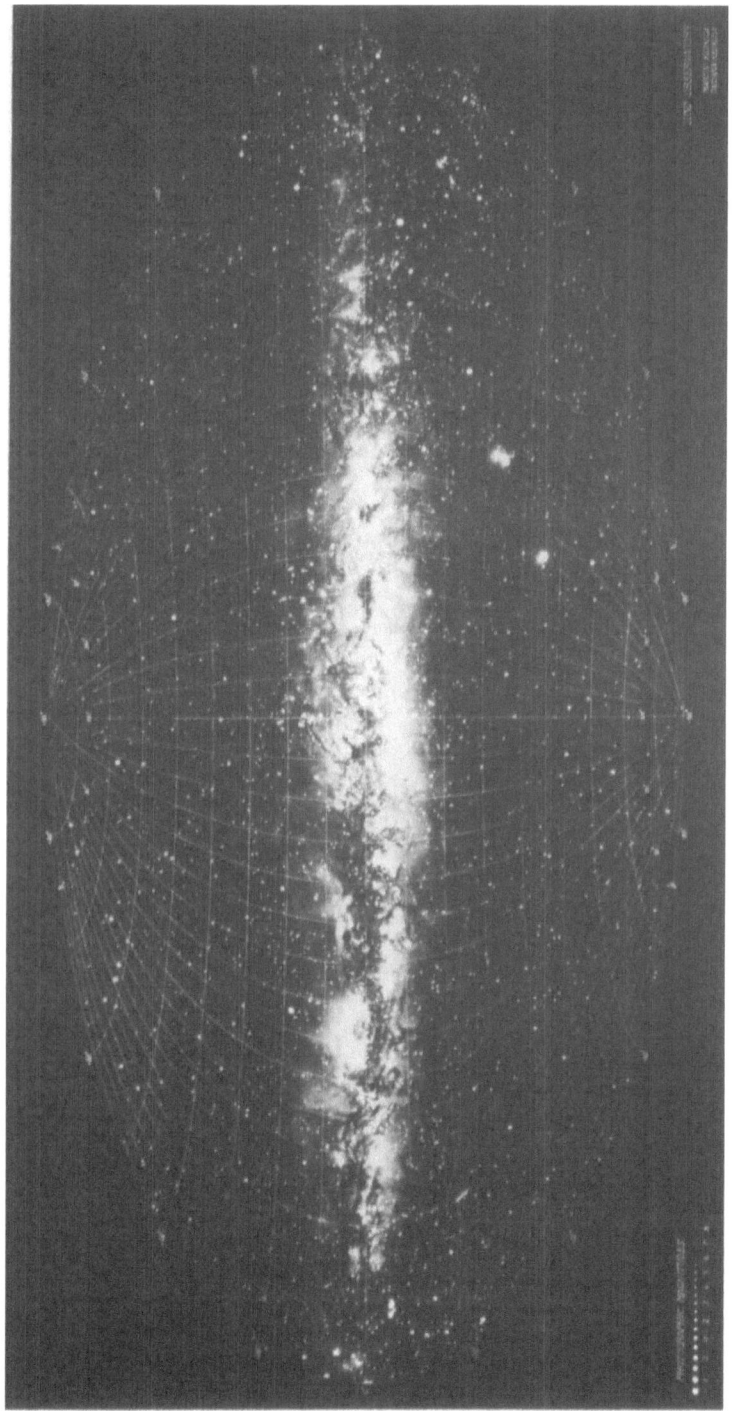

Figure 2   A map of the whole sky in Galactic coordinates. The centre of the Galaxy is in the centre of the diagram. This is a hand-painted map of the sky by M. and T. Keskula of the Lund Observatory.

Figure 3   The Andromeda nebula (M31) , the closest giant spiral galaxy to our own Galaxy at a distance of about 2 million light years . It is probably similar in appearance to our own Galaxy (Courtesy of the Hale Observatories) .

Figure 4   The giant elliptical galaxy M87 . This is the nearest giant elliptical galaxy to our own Galaxy at a distance of about 60 million light years . It lies close to the centre of the Virgo cluster of galaxies (Courtesy of the Hale Observatories).

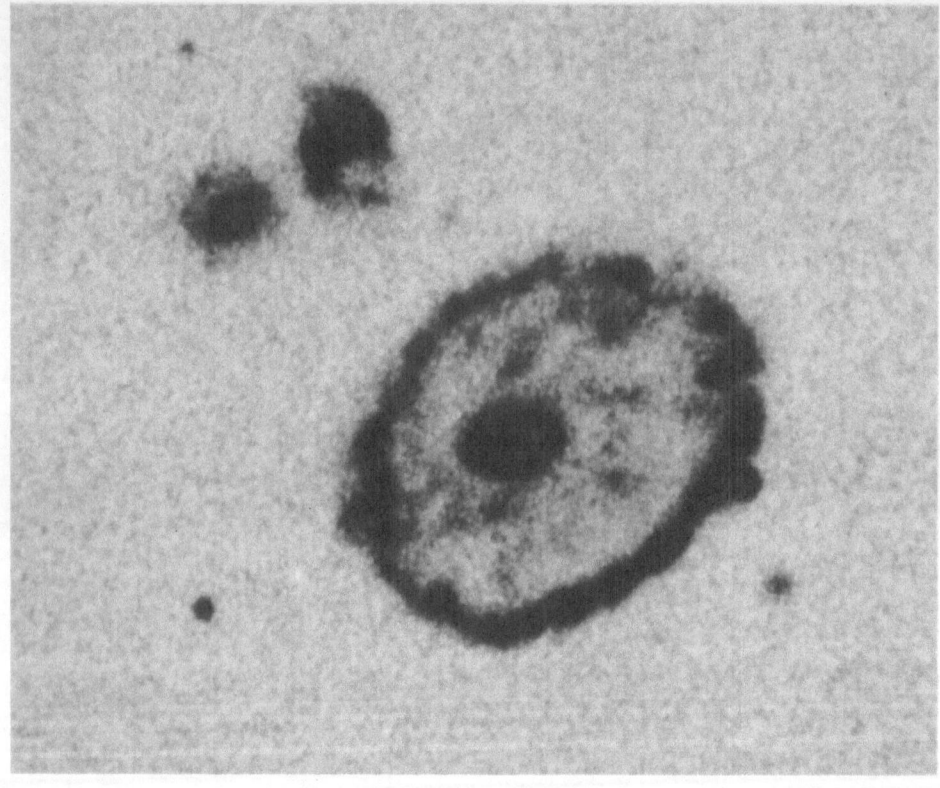

Figure 5   The "Cartwheel" galaxy . A peculiar ring galaxy probably caused by a strong interaction with a nearby galaxy . (Courtesy of the Royal Observatory , Edinburgh) .

To reach the <u>nearest stars</u> , we have to travel a distance about 20,000 times greater than the distance from the Sun to Saturn . This corresponds to a distance of about 4 **light years** . Within a distance of about 50 light years , we know of about 50-100 stars. Many of them are similar to our own Sun but some are fainter and a few brighter .

The stars we see nearby are only a few of the 100,000 million stars which make up our own <u>Galaxy</u> . The Milky Way , which can be seen clearly on a dark night as a bright band across the sky , is the integrated light from all these stars (Figure 2) .   The Milky Way forms what 18th and 19th century astronomers referred to as an "Island Universe" . Namely , the stars which we see locally are not distributed uniformly throughout the Universe but are clustered into giant associations known as galaxies . If we could look at our Galaxy from outside , it would probably look similar to our giant companion galaxy in space , the Andromeda Nebula (Figure 3) . The Galaxy is shaped roughly like a discus with a pronounced central bulge . The Solar System is located towards the edge of the disc on the inner edge of a spiral arm , similar to those seen in spiral galaxies. The size of our Galaxy is about 10,000 times the distance from the Sun to the nearest stars , i.e. a size of about 30,000 light years .

To the cosmologist , galaxies are the building blocks by which the large scale structure of the Universe is defined . They come in a wide variety of different luminosities , shapes and forms . There are spiral galaxies like the Andromeda Nebula , elliptical galaxies like the giant elliptical M87 in the Virgo cluster (Figure 4) and peculiar galaxies like the Cartwheel galaxy (Figure 5) where some dreadful accident seems to have taken place which results in the rapid formation of stars in a ring-like structure .   However , most galaxies in the Universe are "normal" in the sense that they are either spirals or ellipticals . A very small fraction of them , in particular some of the most massive galaxies , are what are called "active galaxies" in the sense that there is violent activity going on in their nuclei and we will discuss their significance later .

Galaxies have a strong tendency to be clustered . This ranges from pairs and small loose groups of galaxies to giant rich clusters of galaxies which are

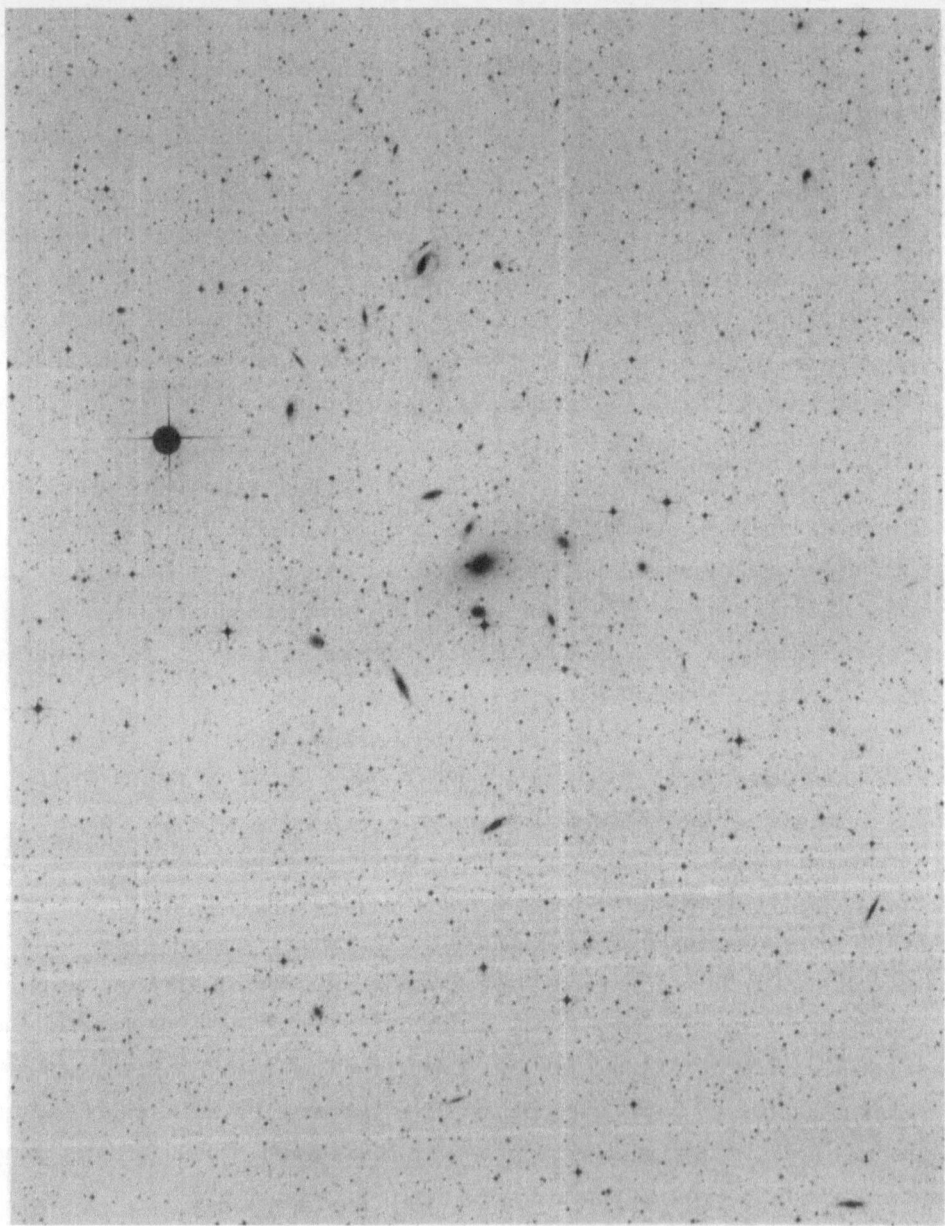

Figure 6   The Pavo cluster of galaxies . A nearby rich cluster of galaxies in
the Southern Hemisphere (Courtesy of the Royal Observatory , Edinburgh) .

among the most striking objects in the sky. Figure 6 shows the nearby Pavo cluster of galaxies where it can be seen that there is a wide variety of galaxies of different types in the same area of sky. A giant rich cluster of galaxies such as Pavo can contain over a thousand galaxies. The typical size of a giant cluster of galaxies is about 100-200 times the size of our own Galaxy. These are rather rare in space and are far outnumbered by smaller groups and clusters which comprise most of the galaxies in the Universe. It is now known that galaxies are very seldom found on their own. One of the basic problems of cosmology is to explain why it is that galaxies are found in such a wide variety of associations.

Going up in scale yet further, we can ask what is the typical distance between these rich clusters of galaxies. In the units, we have been using, the separation between rich clusters is on average about 10-20 times their size. You should picture the distribution of galaxies as being rather clumpy and every now and then coming across a really rich association of galaxies such as the Pavo cluster.

We now know, however, that the distribution of galaxies on this very large scale is more complex than this. Let us look at a picture of the whole northern Galactic hemisphere of the sky with all the stars in our own Galaxy removed. By the Northern Galactic hemisphere, I mean what we see if we look upwards out of the plane of our own Galaxy. We know there is a great deal of obscuration by interstellar dust in the space between the stars in our own Galaxy and therefore we get a highly obscured view of the external Universe if we try to look too close to the plane of the Milky Way.

This large scale map of the Universe is shown in Figure 7. It needs some explanation but is fundamental for studies of cosmology. The coordinate system has been chosen such that the centre of the diagram corresponds to looking vertically out of the plane of our Galaxy and round the edges corresponds to looking through the Galactic plane. The coordinates have been squashed in such a way that if objects were uniformly distributed over this part of the celestial sphere, they would appear to be uniform in this two-dimensional projection. The surface density of galaxies is represented by the brightness at each point on the

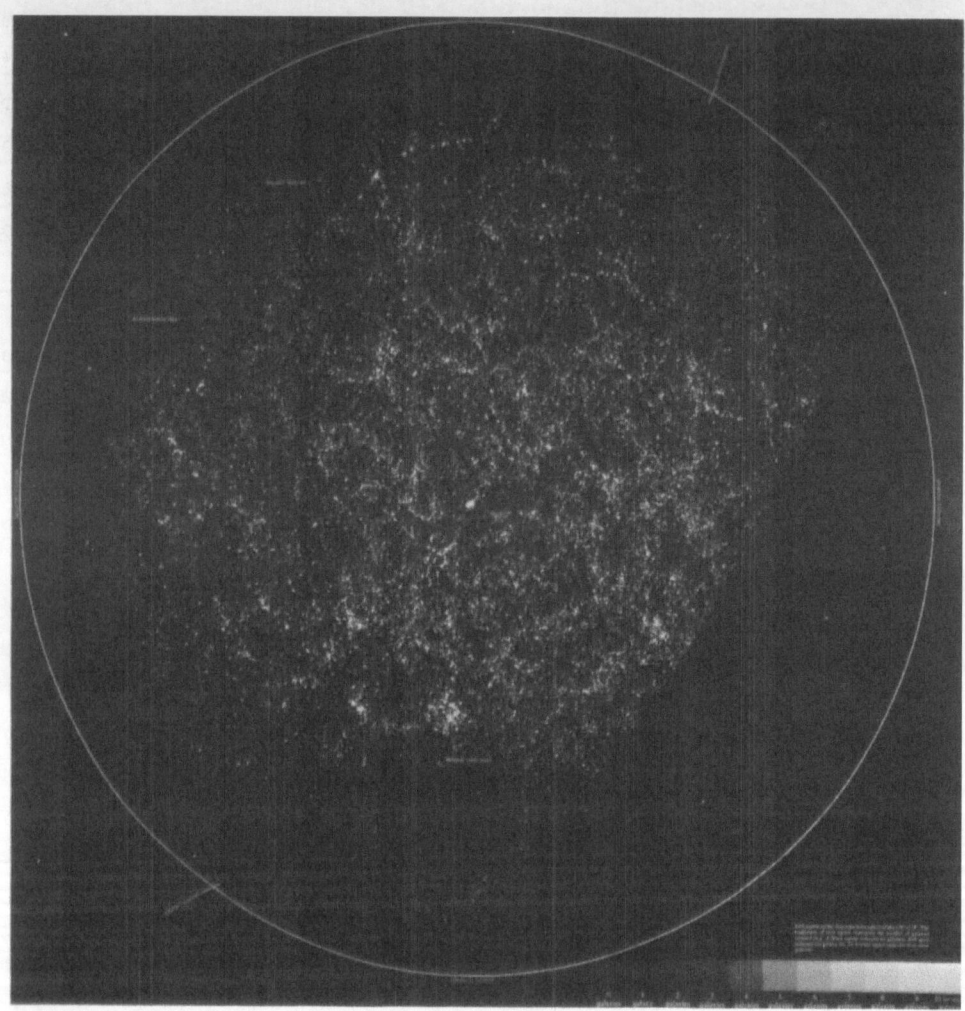

Figure 7   The distribution of galaxies on the celestial sphere out to a distance of about 1,000 million light years derived from the Lick counts of galaxies . The North Galactic Pole is at the centre and the Galactic equator around the circumference. An equal area projection is adopted so that a proper impression is given of the large scale distribution of galaxies . Over 1 million galaxies are represented in this picture .   (From M. Seldner , B. Siebars , E.J. Groth and P.J.E. Peebles , 1977 , Astronomical Journal , 82 , 249) .

celestial sphere , the brighter the region , the larger the number of galaxies . In fact each point on the diagram corresponds to a dimension about the size of a cluster of galaxies .

Two features which are completely artificial are immediately obvious on this diagram . First of all , there is a decrease in the numbers of galaxies towards the edge of the diagram . This is entirely due to the fact that , close to the plane of the Galaxy , the external Universe is seen through the interstellar medium of our Galaxy which , because of the presence of dust , obscures our view . Second , there is a large bite out of the diagram at the bottom right . This is simply because this area of sky was not investigated by the astronomers who counted the galaxies .

Thus , we only get an unobscured view of the Universe when we look towards the central part of Figure 7 . In fact the diagram contains over a million galaxies which were counted by hand by the Lick Observatory astronomers , Shane , Wirtanen and their colleagues . Figure 7 was produced from their counts by Dr P.J.E. Peebles and his colleagues at Princeton University .

The important things to note about this very beautiful picture are as follows . First , the cross-section of the Universe which we see in the diagram corresponds to a view of the Universe at a distance of about 1,000 million light years . In the units we have been using , this is about 4 times the typical distance between giant clusters of galaxies . When we look across the diameter of the diagram , we are looking over a section of the universe about 3,000 million light years across . This is roughly 10 times the distance between giant clusters of galaxies . This is therefore a view of the Universe on the grandest scale .

Second , if we look at the unobscured region towards the centre of the diagram , we note that , if we take averages over large enough regions , the Universe looks more or less the same from one region to another . There is fine scale structure but looked at globally the average numbers of objects do not vary very much from one region to another . This is evidence that if we look on a large enough scale in different directions in the Universe , we get roughly the same picture , i.e. it is evidence for the isotropy of the Universe on a large scale .

Third , despite the large scale uniformity , there is also evidence for real structure on a smaller scale . In fact , it can be seen that on a scale of about one-tenth the diameter of the diagram , there seem to be structures with widely varying densities of galaxies . It is intriguing that this is typically the scale which we associated with the distance between rich clusters of galaxies . Clearly , however , since a single cluster of galaxies would occupy a very small region of the diagram , less than , say , 1/100 of its diameter , the structure seen in Figure 7 cannot be just due to the random positioning of clusters in the Universe . There seems to be a "cell-like" structure in the distribution of galaxies . The galaxies are distributed in clusters , sheets and filaments . One can see features which look like holes in the distribution . It is found that the rich clusters of galaxies are all associated with regions of enhanced galaxy density . They are found in the filaments and around the edges of holes . This strongly suggests that the rich clusters are simply the peaks of a much more complex distribution of galaxies. Some astronomers have compared Figure 7 with what one would observe if one took a section through a sponge .

The reality of this filamentary structure has been confirmed by studies of the three-dimensional structure of the distribution of galaxies which we will discuss later . The explanation of this structure is one of the most important problems for astrophysical cosmology . We will return to the problem in later sections .

We have now reached very large scales in the Universe . How far can we actually see beyond this scale in the Universe ? The best way of giving an unambiguous answer is to ask "How far can we observe the Universe more or less as it is today ?" The problem is that when we look far away , we look back in time to epochs when the Universe was younger than it is now . So the best way of giving an answer to the question of how far away we can observe the Universe is to say that the size of the Universe which we can observe more or less today is about 50 times the distance between giant clusters of galaxies i.e. about 5-10 times the distance of the objects in Figure 7 , or a distance of about 10,000 million light years . It becomes less and less meaningful to talk about greater distances because we are looking so far back in time that the Universe was much younger than it is now and time becomes a much more useful measure of the Universe than distance . Notice that we can see objects "further away" than this "size" of the observable Universe if they are bright enough but , as I have emphasised , it is much more meaningful to think in terms of time rather than distance for these very distant objects , as I will explain in the next section .

Notice how we have come from the distance of the Earth from the Sun to the size of the visible Universe in 6 easy stages , at no stage going up in size by factors larger than 20,000 . All this means is that you should think in relative terms rather than absolute terms . To put it another way , you should think logarithmically about distances in astronomy .

Looked at this way , I always think of the Universe as a rather cosy , homely place . Perhaps I have been working in cosmology too long but I feel very much at home in our tiny "Island Universe" within a much larger system of galaxies . We must now begin to construct models for our Universe .

3 . The Case for the Hot Big–Bang

3.1 The Classical Cosmological Models

To begin construction of models of the Universe , we have to make a fundamental assumption . This is known as the **Cosmological Principle** and can be simply expressed by the statement that "we are not located anywhere special in the Universe" . This is a very important assumption because , if we were located in some very special position , we could not be certain that the observations we make of the Universe would enable us to make statements about the properties of the Universe as a whole. To put it in its most extreme form , if we were in a very special position , the Universe about us might have been so organised to deceive us and lead us to incorrect conclusions about its true nature .

The corollary of this assumption is that we are in a typical position in the Universe . In other words , any observer located anywhere in the Universe looking at the Universe at the same time should see the same large scale properties which we observe. The question therefore is, "What are the relevant large scale properties of the Universe about which we should require all observers to agree?"

I am going to develop this story in terms of a number of basic facts about the Universe and end up with three fundamental problems associated with the current best-buy model for the Universe. To begin our story we need immediately three of these basic facts:

Fact No. 1 Olbers' Paradox We know that it was not **Olbers** who first discovered the paradox with which his name is associated and, to make matters worse, that he used it for incorrect reasoning about the existence of matter between the stars. The paradox results from the simple observation that "the sky is dark at night". Consider what we would expect to happen in an infinite, uniform, static Universe. The problem is illustrated in Figure 8. If such a Universe is filled with stars, then if we look along any line of sight, we will

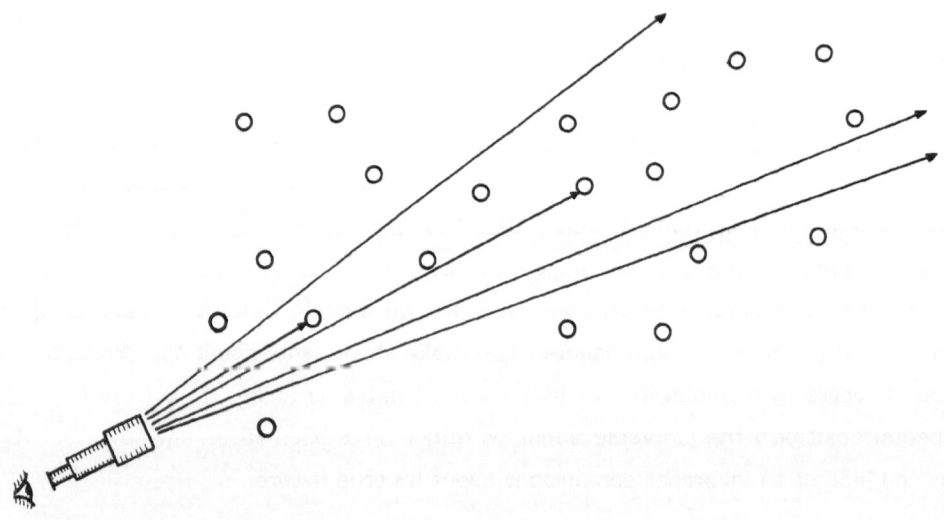

Figure 8   Illustrating the origin of Olbers' paradox.

eventually hit the disc of a star . Therefore , the sky should be as bright as the surface of a star and this plainly conflicts with our experience that the sky is dark at night . In fact , the argument is a very powerful one and the conclusion is that not all three of the assumptions about the Universe can be correct , i.e. it cannot be simultaneously infinite , uniform and static . In fact , **Hubble** showed in 1929 that one of the assumptions is wrong .

Fact No. 2 The Expansion of the Universe What Hubble did was to measure the velocities of galaxies with respect to the Earth . It had been known since 1915 that all the galaxies had velocities which were directed away from our own Galaxy but Hubble's achievement was to show that the more distant the galaxy , the greater the velocity . The important point was that the velocity was linearly proportional to distance . This result has been confirmed by all subsequent observations . A recent demonstration of this relation for the brightest galaxies in clusters is shown in Figure 9 . We can write Hubble's law mathematically $v = H_0 r$ where v is the velocity of the galaxy away from our own Galaxy and r is its distance. H_0 is a constant , known appropriately as **Hubble's constant** , and will turn out to be a measure of the present rate of expansion of the Universe .

It is important to note how Hubble measured the velocities of galaxies . He used the Doppler shifts of lines in their optical spectra . This is the effect with which we are familiar when a police car sounding its siren passes us by . The pitch of the siren drops when its passes us . In physical terms , the wavelengths of the sound waves increase because of the velocity of the sources of the sound away from us . In the case of light waves , we find exactly the same phenomenon . If λ_e is the emitted wavelength of the light by the source and λ_0 the observed wavelength due to the velocity of the source v with respect to us , we can define a redshift of the spectral lines by $z = (\lambda_0 - \lambda_e)/\lambda_e$. It is called a redshift because the lines are shifted towards the red end of the optical spectrum . In the case of velocities much less than that of light , the velocity of the source is just $v = cz$. This is the sort of velocity which Hubble used in discovering the velocity-distance relation . We will return to this point later because in many ways the redshift is much more fundamental than its interpretation as a velocity .

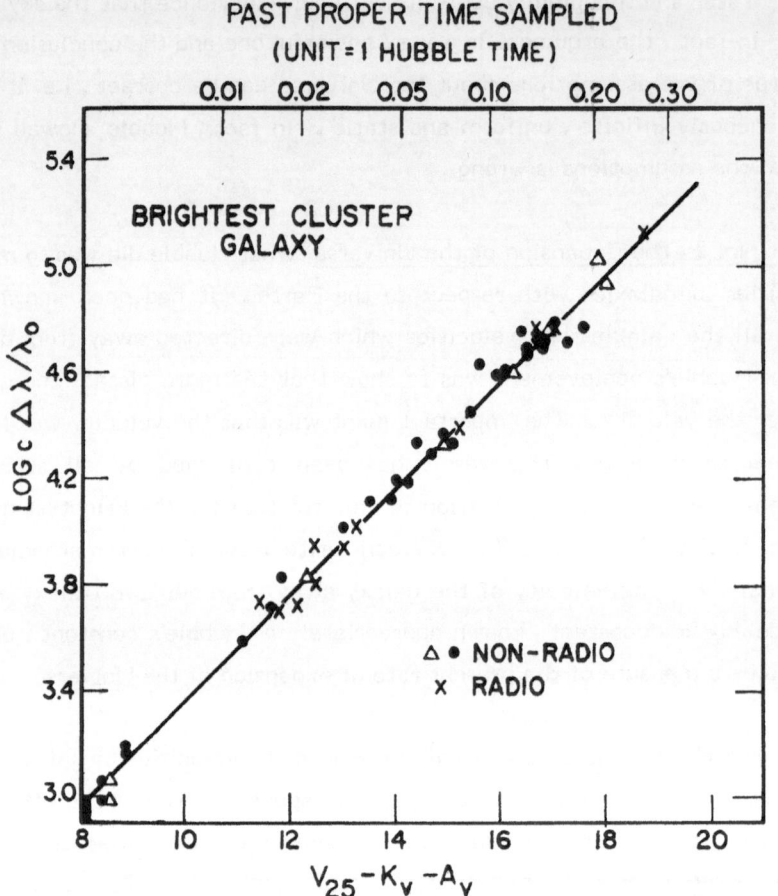

Figure 9   The redshift-apparent magnitude relation for the brightest galaxies in clusters . This diagram is equivalent to the velocity distance relation . This is because (a) redshift can be related to velocity by v = cz at redshifts less than 1 ; and (b) it is found that the brightest galaxies all have roughly the same intrinsic luminosity . Therefore , they can be used as "standard candles" and their apparent brightnesses as measured by their observed intensities (known as apparent magnitudes to astronomers) are a measure of the distance of the galaxy . The straight line shows the relation expected if velocity is proportional to distance (from A.R. Sandage , 1968 , Observatory , 88 , 91) .

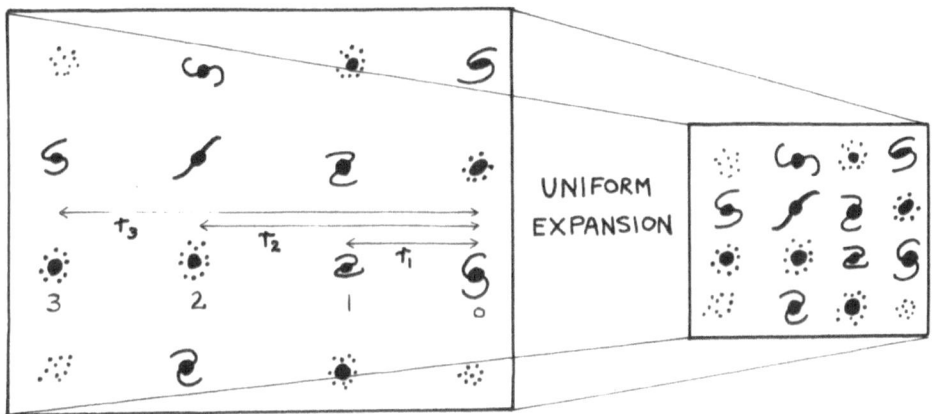

Figure 10   Illustrating how Hubble's law and the isotropy of the distribution of galaxies imply that the Universe as a whole is expanding uniformly .

Now , according to the Cosmological Principle , this result should be found by any observer anywhere in the Universe . We have already seen that the Universe is remarkably isotropic (uniform) and the assumption of isotropy and Hubble's law is sufficient to tell us that the Universe is expanding uniformly as a whole at the present time. This is demonstrated diagramatically in Figure 10 which shows a pattern of galaxies expanding uniformly i.e. in a given time interval , all the distances between neighbouring galaxies in the square pattern increase by the same amount. Now fix your attention upon any one galaxy and measure how far the neighbouring galaxies move in the fixed time interval . You can see that the galaxies numbered 1 , 2 and 3 have moved r_0 , $2r_0$ and $3r_0$ with respect to number 0 , i.e. the greater the distance , the greater the velocity of the galaxy with respect to 0. Notice that this result would apply no matter which galaxy we take as origin .

Thus , Hubble's law tells us that the Universe as a whole is expanding uniformly at the present time .

Fact No. 3   The Microwave Background Radiation   The third basic fact arises from what is probably the greatest discovery in cosmology since that of the expansion of the Universe by Hubble . This is the discovery made in 1965 by **Penzias** and **Wilson** of the Bell Telephone Laboratories of the Microwave Background Radiation . They discovered a strong background signal in their tests of a new high precision antenna which turned out to be the same wherever they looked in the sky .   This was the first detection of a diffuse component of radiation which permeates the whole Universe .   It is present in every cubic centimetre of space at the present time .

Now for our immediate purposes there is only one important aspect of this radiation . Subsequent observations have shown that the microwave background radiation is quite remarkably uniform over the whole sky . In fact , wherever we look in the sky it has the same intensity to better than 1 part in 1,000 (Figure 11) . This is quite amazing precision for any cosmological observation where often one is lucky if one knows anything within a factor of 10 or even 100 in some cases . Thus , no matter what the origin of this radiation , this is very strong evidence that something in the Universe is quite remarkably uniform (or isotropic) and we can use this as a framework within which to build our model of the Universe . Recall that , according to the cosmological principle , we will require all other suitably chosen observers in the Universe to see this highly isotropic radiation as well .

We have now established the two basic properties we need to define this special set of observers who are known as fundamental observers .   They are observers distributed throughout the Universe , all of whom observe the background radiation to be isotropic and who observe the same velocity-distance relation . In other words they are observers who partake in the uniform expansion of an isotropic Universe .

For future reference , I show in Figure 12 the spectrum of the background radiation at all wavelengths from long radio wavelengths to the limit of hard gamma-rays . You should think of this diagram as showing the strength of the background radiation you would see at any point in intergalactic space .   The

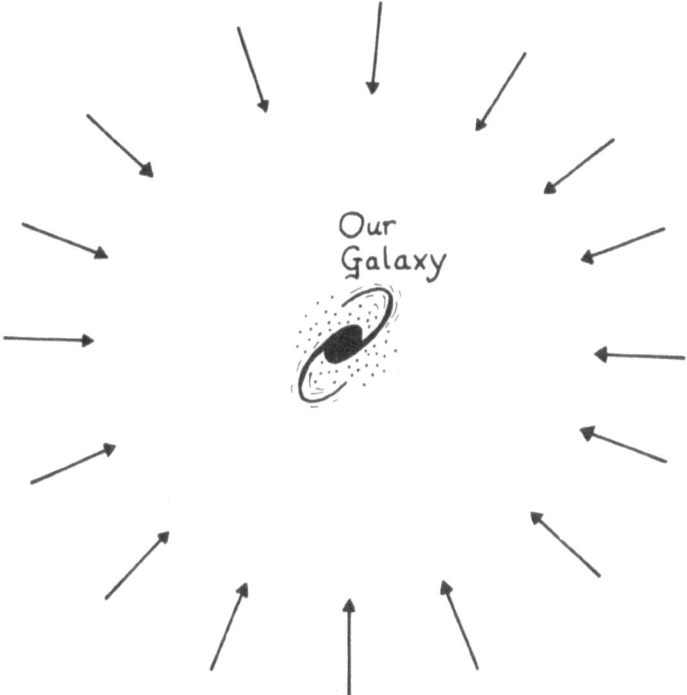

Figure 11   Illustrating the isotropy of the microwave background radiation .

microwave background radiation is the great peak which provides by far the largest contribution to the energy density of radiation in the Universe . We will return to this point again in a moment .

You will also notice in Figure 12 that there is strong background radiation in the X-ray waveband . This is also very important cosmologically . It is also known to be remarkably isotropic , although not with quite the same high precision as the microwave background radiation . This was one of the first discoveries of X-ray astronomy which , as we have discussed , can only be carried out from space vehicles .

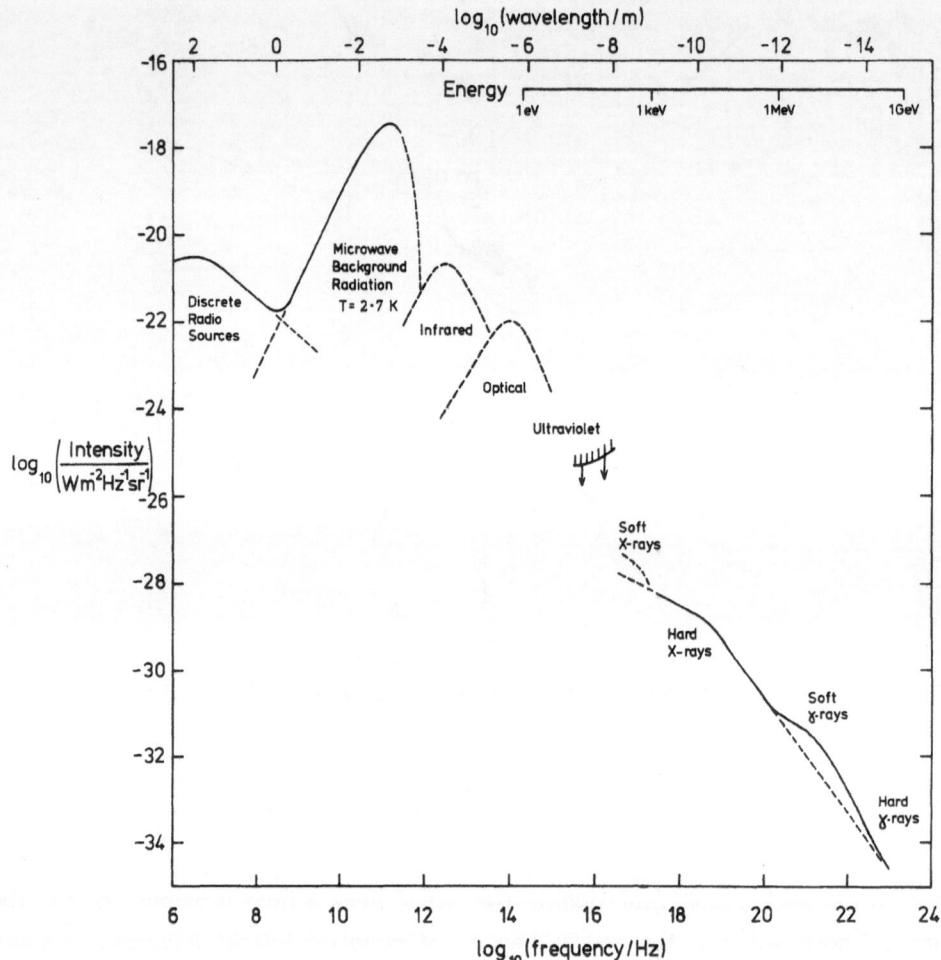

Figure 12   The spectrum of the universal background radiation from long radio waves to the limit of hard gamma-rays . Those portions of the spectrum which have been measured are indicated by solid lines .   Dashed lines indicate theoretical estimates of the intensity of the background radiation . (See e.g. M.S. Longair , 1981 . "High Energy Astrophysics" , p.203 , Cambridge University Press) .

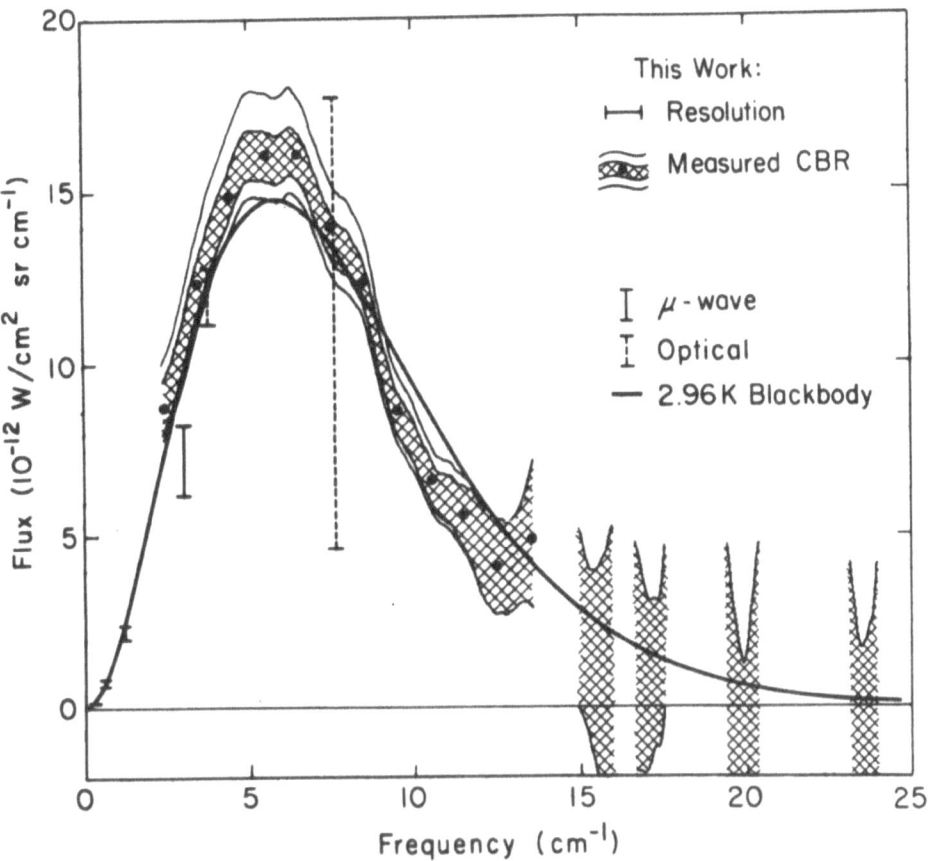

Figure 13   The spectrum of the microwave background radiation measured in a recent balloon experiment by the Berkeley group . The hatched areas indicate the uncertainties in the measurement and the solid line a black body spectrum at radiation temperature 2.9 K .   (From D.P. Woody and P.L. Richards , 1981 , Astrophysical Journal , <u>248</u> , 18) .

Having introduced the microwave background radiation, we should note some of its other properties. Its spectrum is shown in Figure 12 - it has the characteristic form of what is known as black-body radiation or as a Planck spectrum. This is the spectrum of radiation which has had time to come into thermal equilibrium with a hot source of radiation at a particular temperature. For example, if you measure the spectrum of the radiation coming from inside a blast furnace, it has a spectrum of the above form but at a temperature corresponding to that of the furnace. This form of black-body radiation has a very profound significance in physics. It is characteristic of "thermalised" radiation and means that at one time the microwave background radiation must have been in thermal equilibrium with matter at some temperature.

The other remarkable feature of the radiation has already been alluded to - namely, it contributes by far the largest energy density of radiation of any form we know when averaged over the whole Universe. Now, according to **Einstein**, energy is the same thing as mass according to his famous relation $E = mc^2$ and therefore this energy density is the same as a certain amount of mass density in the Universe. The best thing to do is to compare this mass with the average amount of mass in the Universe in forms we know - stars, galaxies, clusters of galaxies and so on. It turns out that the mass density of this radiation is only about one thousandth or one ten thousandth that of the matter density, the uncertainty being due to our uncertain knowledge of average density of ordinary matter in the Universe. Now this ratio may not seem very impressive and indeed it means that we can forget about the amount of mass in the radiation when we work out the dynamics of the Universe at the present day. However, we note that it is the greatest of all the radiation mass densities in the Universe and we will soon see that the picture changes very dramatically when we look at what must have happened to this radiation in the past.

One final point deserves to be made about the microwave background radiation of particular relevance to space science. Figure 13 shows the spectrum of the radiation in the region of the peak of the black-body spectrum and at shorter wavelengths. This is the result of a very beautiful and difficult balloon experiment by Richards and his colleagues from Berkeley. The uncertainties in its exact spectrum are shown by the hatched area . It can be seen that , within the uncertainties , the spectrum is quite a good approximation to that of a black body but one would like to know this with very much greater precision . Indeed , even if there were very small deviations from a perfect black body curve , this would be of profound cosmological importance . At present most of the uncertainties are due to the residual effects of the atmosphere and so the best way of performing this crucial experiment is from a space vehicle . The **Cosmic Background Explorer (COBE)** which , it is planned , will be launched about 1987 could perform this experiment or it could be performed by a Space Shuttle experiment as has been proposed by Professor Derek Martin of Queen Mary College , London . This is an experiment of the greatest importance which we will return to in Section 6.1 .

After this somewhat long digression on the microwave backgound radiation , we get back to our main theme which is what we need to build world models and our first three facts give us a clear indication of how to begin - we should start with world models which are isotropic , uniform , homogeneous and expanding . These models do not contain any of the actual objects we see in the Universe - they are idealised , structureless Universes .

We now need to identify the forces which act on matter in bulk in such Universes . We know that the only long-range force which acts on all matter is **gravity** and therefore another obvious starting point is Einstein's General Theory of Relativity which is the best theory of gravity we have . It is very difficult to test relativistic theories of gravity with high precision because virtually everywhere we observe in the universe , the gravitational fields are weak . Nonetheless , we now know that General Relativity is correct at least at the 1% level from observations within the solar system and , more remarkably , from observations of a recently discovered pulsating radio source (a pulsar) in a close binary system .

We should note that it is an assumption that gravity as described by General Relativity is the dominant long-range force in the Universe . It could be that other forces are important which could only be detected on the largest scale in

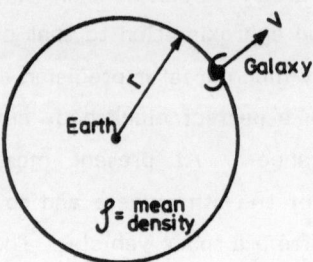

Figure 14 Illustrating a simple model for working out the dynamics of Newtonian Universes .

the Universe and then we would have to revise the models completely . However , the picture which emerges has had some considerable successes which suggest to me that we are basically working along the correct lines .

We can build a very accurate model for the Universe using simple Newtonian gravity . Suppose we consider ourselves to be at the centre of a uniformly expanding sphere of gas and we ask "What are the forces acting on a galaxy at the circumference of this sphere ?" (Figure 14) . The galaxy has radial velocity v and mass m_g . Then according to Gauss' theorem , because the matter is spherically symmetrically distributed about us we can replace the sphere by a point mass of the same mass at the origin , i.e. by a mass $M = \frac{4\pi}{3} \rho r^3$ where ρ is the density of matter in the sphere and r its radius . Now write down Newton's law of gravity for the force acting on the galaxy

$$m_g \ddot{r} = -\frac{GMm_g}{r^2} = -\frac{4\pi}{3} G\rho \, m_g \, r$$

where $\ddot{r}$ is the acceleration of the galaxy . Now we notice that m_g cancels on both sides of the equation . All trace of the galaxy disappears from circulation !

$$\ddot{r} = -\frac{4\pi}{3} \, G\rho \, r \qquad\qquad (1)$$

What has happened is that the equation tells us something in general about the dynamics of a uniform medium under gravity and it applies to the distance r between any two points in the Universe . Indeed , because of the assumptions of isotropy and homogeneity and the fact that every observer who sees an isotropic Universe should also see a uniformly expanding Universe , every observer would perform the same calculation and we need not worry about boundary conditions . The reason this sum works is because in these idealised Universes , local physics is also global physics since every cubic centimetre must have the same properties at the same time .

It is for exactly the same reasons that we have no problem in defining a **cosmic time** . Cosmic time is just measured by a fundamental observer and since they are all equivalent to one another , there is no problem about synchronising their clocks . We often refer to the time from the beginning of the model in terms of various epochs so that I call the present day , the present epoch and refer to it as t_0 .

Now for completeness , let us make two simple substitutions . Let us remove distances completely from the calculation by writing in place of r a scale factor R such that if r_0 is the distance of a galaxy from us at the present epoch in the Universe , the distance at any other epoch would be r = Rr_0 . In other words , R just measures the relative distance between any two points in the Universe as a function of time and is equal to 1 at the present time . Second , let us write the density of the Universe in terms of what is known as the **critical density** at the present epoch $\rho_c = 3H_0^2/8\pi G$ where H_0 is Hubble's constant. ρ is at present the density at any epoch in the Universe . Let us refer all densities to the present average density of the Universe ρ_0 and therefore write $\rho = \rho_0/R^3$. Thus , we can write $\rho_0 = \Omega\rho_c = 3H_0^2\Omega/8\pi G$ where Ω is called the density parameter. The sense in which ρ_c is critical will become apparent in a moment . Equation (1) then becomes

$$\ddot{R} = -\frac{1}{2}\frac{\Omega H_0^2}{R^2} \qquad\qquad (2)$$

It is remarkable , but wholly understandable , that a full detailed treatment using General Relativity results in exactly the same simple equation .

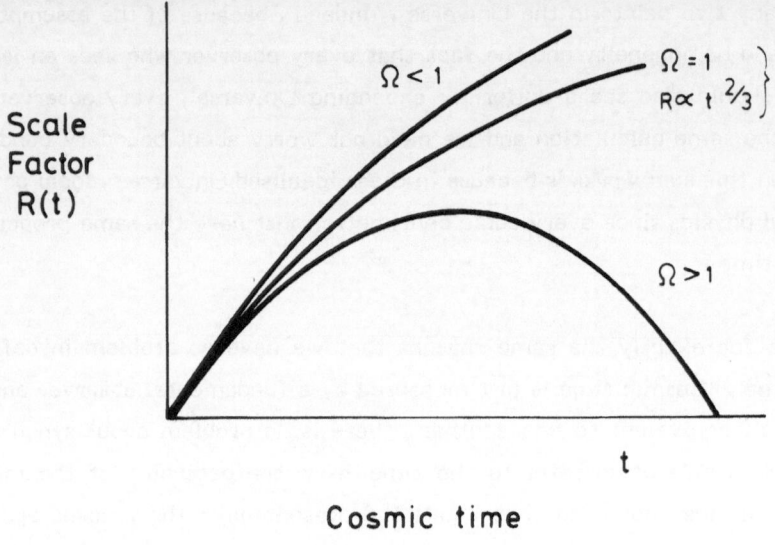

Figure 15   The variation of the scale factor R as a function of time for the classical world models (see also Table 2) .

Now what we need are the answers and these can be readily found from elementary mathematics . The solutions of equation (2) are shown in Figure 15 and listed in Table 2 . Three basic types of solution are found:

a)   There are universes which expand forever and end up at infinity with a finite velocity ; these have density parameter $\Omega = \rho_o / \rho_c$ less than 1 .

b)   there are Universes which expand to a maximum size and collapse again ; these have densities greater than the critical value $\rho_o > \rho_c$.

c)   there are perfectly balanced models which just reach infinity at infinite time with zero velocity . This is the critical model with density $\rho_o = \rho_c$.

TABLE 2

THE PROPERTIES OF CLASSICAL WORLD MODELS

Density $\Omega = \dfrac{8\pi G\rho_0}{3\,H_0^{\,2}}$	Deceleration Parameter $q_0 = -\,(\ddot{R}/R^2)\,t{=}t_0$	Dynamics of Universe	Geometry
$\Omega < 1$	$q_0 < \tfrac{1}{2}$	Expands to infinity	Open , hyperbolic geometry , $k = -1.$
$\Omega = 1$	$q_0 = \tfrac{1}{2}$	Just expands to infinity at infinite time	Flat Euclidean geometry $k = 0$
$\Omega > 1$	$q_0 > \tfrac{1}{2}$	Universe expands to maximum size and then collapses	Close spherical geometry $k = +1$

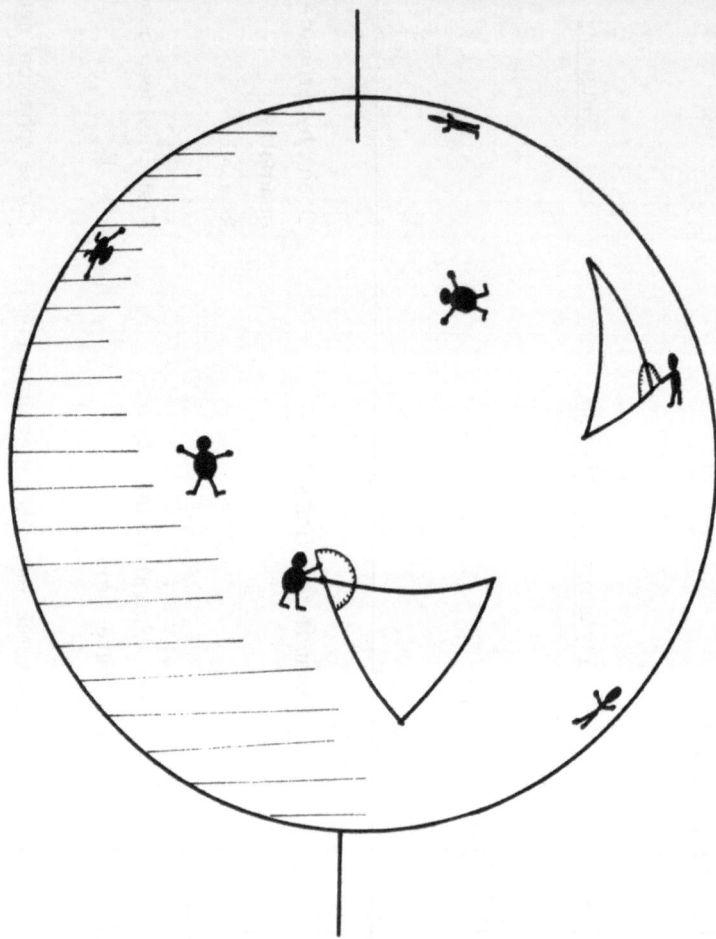

Figure 16   Illustrating two-dimensional people measuring triangles in an isotropic non-Euclidean two dimensional Universe. The topology of the Universe is spherical and the geometry closed .

This behaviour can be understood in terms of the "escape velocity" of the Universe in the same sense as the escape velocity of space vehicles from the Earth . It all depends on the initial velocity with which the system began . If the velocity were large enough to begin with , the decelerating effect of gravity will not be able to prevent the spaceship (or the galaxy) escaping to infinity . If the velocity were not great enough , the spaceship falls back to the Earth . In the cosmic analogy , there is so much matter in the Universe that the galaxy is decelerated and eventually falls back again . The $\Omega = 1$ model corresponds to the spaceship having exactly the escape velocity .

Another miracle of the General Relativistic models of the Universe , which cannot come out of the Newtonian model is the spatial curvature of the Universe . We are very used to thinking in terms of flat 3-dimensional Euclidean geometry which is adequate for all terrestrial purposes . However , there are other geometries which satisfy the requirements that they should be the same at all points in an isotropic Universe . In a two dimensional Universe , for example , we could have a spherical geometry in which two dimensional surfaces are closed as on the surface of a sphere . Two dimensional people who can only see their Universe within the two dimensional surface could measure the angles of a triangle and would find that they get the same non-Euclidean answer , i.e. greater than 180° , everywhere on the surface for a triangle of a given size (Figure 16) . Thus , the surface of a sphere is an isotropic , non-Euclidean space for 2-dimensional people .

The miracle of General Relativity is that there is one-to-one relation between the geometry of the Universe on a global scale and its mean density. This is listed in Table 2 . In principle , the curvature of space is a locally measurable quantity ; we could measure the angles of a triangle with extreme precision and find out if they add up to 180° . However , the effect is totally unmeasureable locally .

There are several possible tests of this picture but all of them are very difficult . We have already introduced H_0 , Hubble's constant and Ω the density parameter . We can readily show that Hubble's constant H_0 is just a measure of

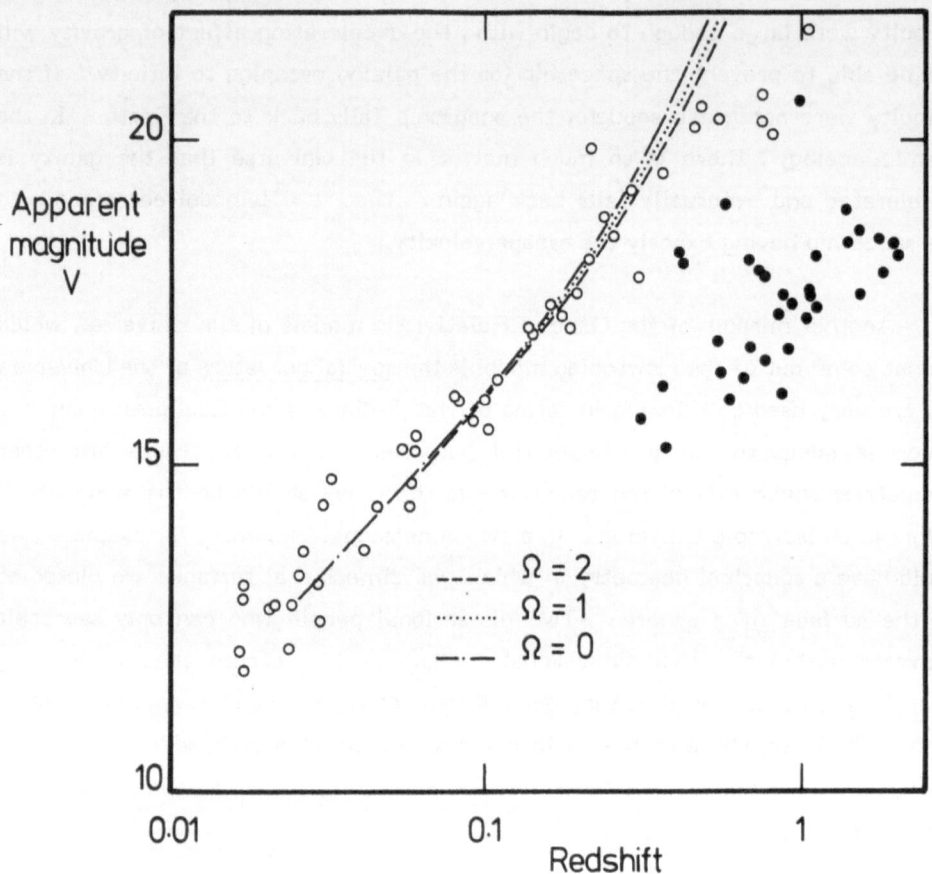

Figure 17   The redshift-apparent magnitude relation for radio galaxies and quasars . The radio galaxies are indicated by open circles and the quasars by filled circles . The expected relation for radio galaxies is shown by the solid and dashed lines for three different values of Ω . It can be seen that the open circles tend to be below the predicted relation at large redshifts .

the rate of expansion of the Universe as a whole at the present day, i.e. mathematically, $H_0 = \dot{R}_{t=t_0}$. Another measurable parameter is the present deceleration of the Universe, R : we measure this by measuring the velocity of expansion of the Universe in the past and at the present day. Another miracle of the classical world models of General Relativity is that according to these models, if we measure the deceleration by the dimensionless quantity $q_0 = -(\ddot{R}/\dot{R}^2)_{t=t_0} = -\ddot{R}/H_0^2$ measured at the present epoch, we must have $q_0 = \frac{1}{2}\Omega$, i.e. the deceleration is wholly determined by the amount of mass present in the Universe. Since, in principle, we can measure q_0 and Ω separately, we should find the above relation to be correct if the General Relativistic models are correct. Unfortunately it is very difficult to measure q_0 and Ω separately, let alone compare them. What we do know is that they cannot be very different from each other. A basic problem is that we can only measure Ω by dynamical means using the dynamics of galaxies and it might be that most of the mass in the Universe is in some invisible form which is not distributed like the galaxies.

In the light of this problem, many astronomers assume General Relativity to be correct and then have one fewer basic constant to measure for the Universe. The classical cosmologist hopes to determine the value of q_0 from observations of distant galaxies. That the observed properties of the galaxy should depend upon q_0 is clear because, if q_0 is large, i.e. there has been considerable deceleration of the Universe, the galaxy will not be as far away when it emitted its light as in one of low q_0. Thus, in Universes with larger q_0, galaxies of the same type should be brighter than those in low density Universes. Of course, this naive argument has to be dressed up in the formalism of isotropic curved spaces.

It is not appropriate to go into the details of the great problems of determining H_0, q_0 and Ω here. Let us note that Hubble's constant H_0 lies in the range of about 50 to 100 km s^{-1} Mpc^{-1}, $\sim 1.6 - 3.3 \times 10^{-18}$ s^{-1}, my own preference being for values towards the lower end of that range. This is because the maximum age of the Universe according to the classical models is one in which no deceleration has taken place and therefore the age must be less than

H_0^{-1}. With $H_0 = 50 \text{ km s}^{-1} \text{ Mpc}^{-1}$, $t_{cosmological} = H_0^{-1} \sim 20 \times 10^9$ years which is just greater than the ages of the oldest known systems in our Galaxy, the globular clusters, which have ages $\sim 17 \times 10^9$ years. If the Hubble constant were $100 \text{ km s}^{-1} \text{ Mpc}^{-1}$, the classical world models would have ages less than 10×10^9 years and then we would have to find something drastically wrong with the theory of the evolution of stars in globular clusters or we would have to adopt some new physics for the large scale dynamics of the Universe.

The value q_0 is even more uncertain. The relation between intrinsic properties and observed properties of distant objects depends upon q_0 (or Ω). There are therefore two approaches which can be taken. Either : we must have such confidence in our understanding of the astrophysics of galaxies, radio sources, quasars, etc., that we can derive their intrinsic properties independent of distance. This requires a high degree of sophistication in understanding the physics of these distant objects and this is lacking for all classes of object. Or : we make the assumption that the physical properties of similar objects nearby and far away are the same. Then, if objects of the same physical types can be identified, the variation of observed properties with redshift may be used to find q_0.

The basic problem with the latter approach is finding suitable "**standard candles**" or "**rigid rods**" among the properties of distant galaxies at any wavelength. Even worse is the fact that we cannot be sure that the "standard" properties have not changed with cosmic epoch. Since we have to look back to epochs when the Universe was significantly younger than it is now, there is no a priori reason why the physical properties of galaxies should not have changed. In Figure 17, for example, the redshift-apparent magnitude relation for radio galaxies and quasars is shown. Although the galaxies follow the expected relation at small redshifts, they deviate from all the predictions at large redshift. We now understand that at least part of this is due to the fact that these galaxies were more luminous in the past. Therefore, it is necessary to understand the evolution of these galaxies with cosmic epoch before an estimate of q_0 can be made. I would not like to quote any value for q_0, except to note that it is unlikely to be very much greater than 1.

In many ways measuring Ω , the density parameter , is easier . All methods of measuring masses in astronomy are dynamical . For example , the masses of galaxies are found from measuring their rotational velocities and the masses of clusters of galaxies from the velocity dispersion of the galaxies of clusters . A method of setting a lower limit to Ω is to add up the mass in all the visible matter in galaxies . Most astronomers agree that the value obtained in this way is about $\Omega = 0.01$. Another attractive technique is to measure the velocities of galaxies close to very large scale systems such as superclusters . Their trajectories should be significantly affected by the presence of such a large mass . On the very largest scales , one can use versions of what is known as the **"cosmic virial theorem"** which essentially compares the relative kinetic and potential energies of the fluctuating component of the mass distribution in the Universe . These last estimates give values of $\Omega \sim 0.1\text{-}0.3$. Thus , it seems that these observations suggest we live in an open , ever expanding Universe but I do not believe any astronomers consider the matter settled . My own view is that probably Ω and q_0· will fall out of some new physical argument about the properties of distant objects . Most of us would much prefer to use a physical argument than one which relies upon individual objects in the Universe being well behaved .

3.2 Fact Number 4 The Hot Big-Bang Model of the Universe

I have sufficient confidence in the argument which follows that I have elevated the Hot Big-Bang Model of the Universe to the status of a " **Fact** " . The argument runs as follows :

So far we have considered only the matter content of the Universe but what about the radiation ? This is where the microwave background radiation is of singular importance . What happens when we squash radiation in the Universe ? The answer is illustrated in Figure 18 . We can simulate the Universe by a box with perfectly reflecting walls and then ask how the wavelength of radiation changes if we expand the size of the box . The answer is that the wavelength just expands with the size of the box . If we write λ for wavelength , we find $\lambda \propto R$.

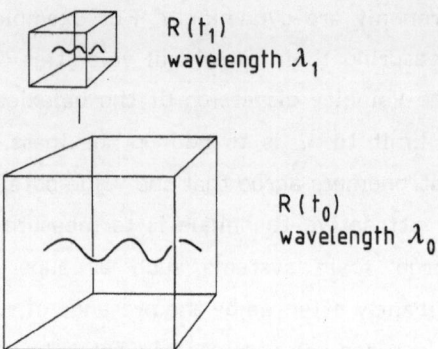

The Universe as an expanding box with perfectly reflecting walls.
The scale factor R(t) measures the relative size of the Universe at time t.

$$\frac{\lambda_1}{\lambda_0} = \frac{R(t_1)}{R(t_0)}$$

Figure 18   Illustrating the way in which the wavelength of radiation changes in an expanding Universe .

This is in fact the fundamental meaning of redshift in cosmology . If the galaxy emitted radiation of wavelength λ_e and it is received at the present epoch with larger wavelength λ_0 , the redshift z of the galaxy is defined as

$$z = \frac{\lambda_0 - \lambda_e}{\lambda_e}$$

However , we have shown that $\lambda_e / \lambda_0 = R$ and hence $R = 1/(1+z)$. Optical astronomers measure the shift of the spectral lines and interpret it as a Doppler velocity of the source away from the Earth . For velocities much less than the velocity of light , $v = cz$ where c is the velocity of light .   These are the velocities used in the velocity-distance relation which defines Hubble's law . The present argument shows that by measuring the redshift z of a galaxy , we in fact measure the size of the Universe when the galaxy emitted its radiation . $R = 1/(1+z)$. This is a much more fundamental interpretation of the redshift .

We can easily show that black-body radiation preserves its characteristic Planckian spectrum (Figures 12 and 13) in expanding Universes but corresponds to a radiation temperature $T = T_0/R = T_0(1+z)$ where T_0 is the present radiation temperature which is about 2.9 K . Even more important is the fact that the total energy in the radiation increases into the past . The matter density increases as R^{-3} as we squash the box but the energy (or mass) density in the radiation increases as R^{-4} ; three powers of R arise because the size of the box is smaller and one power because the energy of each wave is increased by a factor R^{-1}. This last factor arises because the energy of each wave is inversely proportional to its wavelength , $E \propto \lambda^{-1}$.   Thus , in addition to the background radiation temperature increasing , the amount of mass in the radiation increases relative to the amount of ordinary matter as R^{-1} . This means that if we squash the box by a factor of about 10,000 , there must be at least as much mass in the radiation as there is in the matter because we know that at present the relative amount of mass in matter and radiation is 1,000 or 10,000 to 1 .

What does this mean in dynamical terms ? We have already proved the formula for the dynamics of a Universe consisting only of matter :

$$\ddot{R} = -\frac{4\pi G\rho_0}{3}\frac{1}{R^2}$$

ρ_0 is the present density of matter .   What we have shown is that when R becomes very small , the mass density associated with radiation ρ_{rad} which has a different dependence on R . i.e . $\rho_0 \propto R^{-3}$ whereas $\rho_{rad} \propto R^{-4}$. All that happens is that the dependence of $\ddot{R}$ upon R changes from R^{-2} to R^{-3} and the dynamics of this "radiation dominated" Universe are simple , $R \propto t^{\frac{1}{2}}$. This is what cosmologists mean by the distinction between a <u>matter</u> and <u>radiation-dominated Universe</u> .   At late epochs $R > 10^{-3}$, the matter content of the Universe defines the dynamics ; at earlier epochs the radiation is more massive than the matter and dominates the dynamical evolution of the Universe .

We can now put all of this together to work out the temperature history of the Universe . In Table 3 , we list various significant epochs in the history of the

Universe . These deserve some comments . The Table lists the scale factor and tells us by how much we squash the Universe , the time (or epoch) when this takes place after the big-bang , the significant events , the temperature of the thermal background radiation and the density of the matter component in the Universe , assuming the value at the present day to be about 10^{-30} g cm^{-3} which corresponds roughly to $\Omega \sim 0.3$.

The epoch of "recombination"   When the background radiation temperature rises to about 4000K , there is sufficient short wavelength radiation present in the Universe to <u>ionise</u> all the neutral hydrogen . Now most of the matter in the Universe is in the form of the lightest element , hydrogen . By ionise , we mean separate the electron from the proton in the hydrogen atom , in this case by shining ultraviolet light on the gas . This means that at earlier times there is no neutral hydrogen - it forms instead a fully ionised <u>plasma</u> , the phase of matter which is found in the ionosphere of the Earth , in the Sun and in plasma physics experiments . The epoch is called the recombination epoch because when we run the clocks forward , the hydrogen plasma begins to recombine to form a neutral hydrogen gas at this time .

We have already discussed the slightly earlier epoch when there were equal amounts of mass density in the matter and radiation . It is coincidental that this happens to take place at roughly the same epoch at which recombination takes place . Another interesting physical feature of the evolution of the plasma at epochs earlier than this is that as soon as the hydrogen is all ionised , there is very strong thermal coupling between the radiation and the plasma . This takes place because , whereas neutral gas does not scatter radiation , the free electrons scatter the radiation very strongly and thus can transfer energy between the matter and radiation .   This efficient scattering process , called <u>Compton scattering</u> , ensures that the radiation and matter remain at the same temperature at all epochs prior to the epoch of recombination .

Epoch of nuclear reactions   When the Universe is squashed by a factor of 1,000 million , the temperature of the plasma is about 3,000 million degrees and the thermal spectrum has maximum intensity at gamma-ray energies .   These

TABLE 3

SIGNIFICANT EPOCHS IN THE UNIVERSE

Scale Factor R	Time from Big Bang	Events	Temperature of Radiation	Matter density ($g\ cm^{-3}$)
1	2×10^{10} years	Now	3K	10^{-30}
1/1500	10^7 years	At this temperature all the neutral hydrogen in the Universe is ionised	4000K	10^{-20}
1/1000 – 1/10,000	2-20 $\times 10^6$ years	There are equal amounts of mass in the matter and radiation. At earlier times the Universe is radiation dominated	3,000 – 30,000K	10^{-21} – 10^{-18}
1/10^9	10 minutes	The radiation is so hot that the nuclei of atoms dissociate	3×10^9 K	10^{-3}
1/3 $\times 10^9$	1 minute	Electron positron pairs are created from the thermal background radiation	10^{10}K	0.03
1/10^{13}	$\sim 10^{-5}$ seconds	Proton-antiproton, baryon-antibaryon pair production from the thermal background radiation	$\sim 10^{13}$K	10^9

waves are so energetic that they can dissociate the neutrons and protons which make up the nuclei of atoms . We can therefore be certain that , at earlier epochs , there will be no atomic nuclei as we know them - they will all be broken down into their constituent parts . In other words , at this epoch , the Universe consists of protons , neutrons , electrons , photons and various forms of neutrino which accompanied earlier interactions .

Electron-positron pair formation   If the Universe is squeezed just a little further , the gamma-rays become so energetic that they can collide together and form electron-positron pairs . The positron is the anti-particle of the electron and is well known in laboratory experiments . The basic physical principle involved is that when the gamma-rays which collide have total energy greater than twice the rest-mass energy of an electron ($E = m_e c^2$) , there is a probability that the pair of gamma-rays will be transmuted into an electron and its antiparticle , the positron. This can be written schematically

$$\gamma + \gamma \rightarrow e^+ + e^-$$

At this temperature of about 10,000 million degrees , this process becomes feasible and a new equilibrium is set up in which there are roughly equal numbers of electrons , positrons and gamma-rays .

We may well ask , "How can we be sure that we understand the physics at such high densities and temperatures ?" The answer to this is very interesting : the last column in Table 3 shows that , although the temperatures are very high , the density of matter is very modest . In fact , when the scale factor has value $1/3 \times 10^9$, the mass density of ordinary matter is only about 0.03 g cm^{-3} which is 30 times less than the density of water . Thus , although we are talking about high temperatures , the densities are still at values which are similar to those found in terrestrial laboratories .

Proton-antiproton pair production   We can go further back still and ask what happened when the scale factor was only one ten million millionth of its present value ($R \sim 10^{-13}$) . Then the temperature is so high that the gamma-rays can collide and create protons and their antiparticles , the antiprotons . This is exactly the same process as electron-positron pair production described above but now because the rest mass of the proton is 1,800 times greater than that of the electron , proton-antiproton pair production takes place at 1,800 times higher temperature. In fact above this temperature pair production of all the heavy particles known to elementary particle physicists can now take place . These particles are generally known as baryons (meaning "heavy particles") and in general the process is referred to as baryon-antibaryon pair production . Now the densities are becoming very high and soon one approaches the densities found in the nuclei of atoms .

At earlier epochs , we begin to run out of secure physics . Up to this point we have been using "known" physics in the sense that the nuclear interactions have been studied in terrestrial accelerators . At higher energies , the nuclear physics is not yet available on an experimental basis , although many elementary particle physicists have an excellent picture of what they can expect to happen . We can be certain that at these very early epochs , all the different types of elementary particles discovered by particle physicists come into equilibrium . The Universe at these very early times bore very little resemblance to our present Universe. Notice that the time-scales we are talking about are now rather short . They are listed in Table 3 as time from the Big Bang at t = 0 .

This summary of what we expect to happen in conventional world models makes it abundantly clear why they are known as Hot Big-Bang Models of the Universe . We should note immediately some of its great successes . First , it is entirely natural that the microwave background radiation should be isotropic . Second , it is natural that the microwave background radiation should have a pure black-body spectrum . In the very early Universe , all the constituents are in thermal equilibrium and so the radiation spectrum should take up the equilibrium form which is that of black-body radiation . After the early stages , the thermal radiation cools as we have described above but preserves its thermal radiation spectrum .

This is an attractive picture but it would be reassuring to have some completely independent evidence for these hot early stages of the Universe . Fortunately , we now have this evidence by a quite remarkable argument . Let us

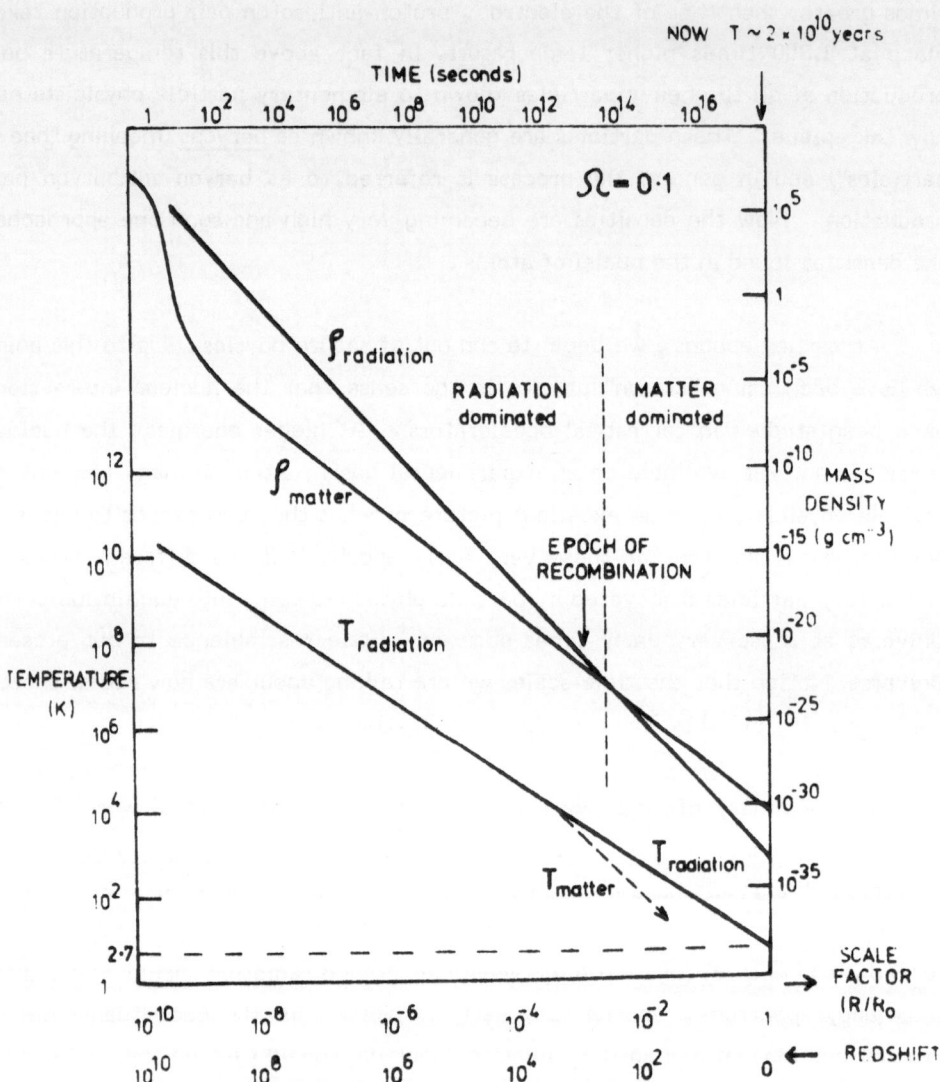

Figure 19   The temperature history of the Universe . The diagram summarises
the expected variation of the radiation temperature and density matter with
cosmic epoch for a low density Universe , $\Omega = 0.1$.

run the clocks forward and see how we would expect the various constituents of the Universe to evolve with time . This is illustrated in Figure 19 which summarises the various phases described above .

It has always been a problem in astronomy to understand the origin of some of the lightest elements . We are confident that heavy elements like Carbon , Nitrogen , Oxygen , Iron and so on are produced by nuclear reactions in the centres of stars . There are a number of separate processes which are responsible for different elements but in broad terms we can understand where they came from. The main problem lies with elements such as the isotopes of Helium , Helium-3 (^{3}He) and Helium-4 (^{4}He) , Deuterium D and Lithium-7 (^{7}Li) . The basic problem is that these are all rather fragile elements and if they are mixed into the hot inner regions of stars , they are rapidly destroyed or converted into heavier species .

The mystery is deepened by the fact that these light elements seem to have more or less the same abundance by mass wherever they are observed . Wherever it is possible to observe Helium-4 , it always turns out to have an abundance of about 23% or greater by mass relative to Hydrogen . This is much greater than could ever be generated in stars .

Perhaps the most important contribution of space science to cosmology up till now has been the detection of ultraviolet absorption lines of deuterium (D) in the interstellar gas of our own Galaxy . Deuterium does not possess any useful lines in the visible part of the spectrum but in the far ultraviolet region , which is only accessible from a space observatory , absorption lines similar to those of hydrogen can be observed . Not only has it been possible to measure the deuterium to hydrogen ratio in the interstellar gas , it has also been possible to measure the relative abundances of the elements along a number of different lines of sight through the gas . Along these different directions , there may be variations in the abundances of the heavy elements , but the deuterium-to-hydrogen ratio always remains more or less the same . In fact , virtually everywhere the D/H ratio is about 1.5×10^{-5} by mass . Although this may seem a low value , it is enormously greater than would ever be expected from stellar

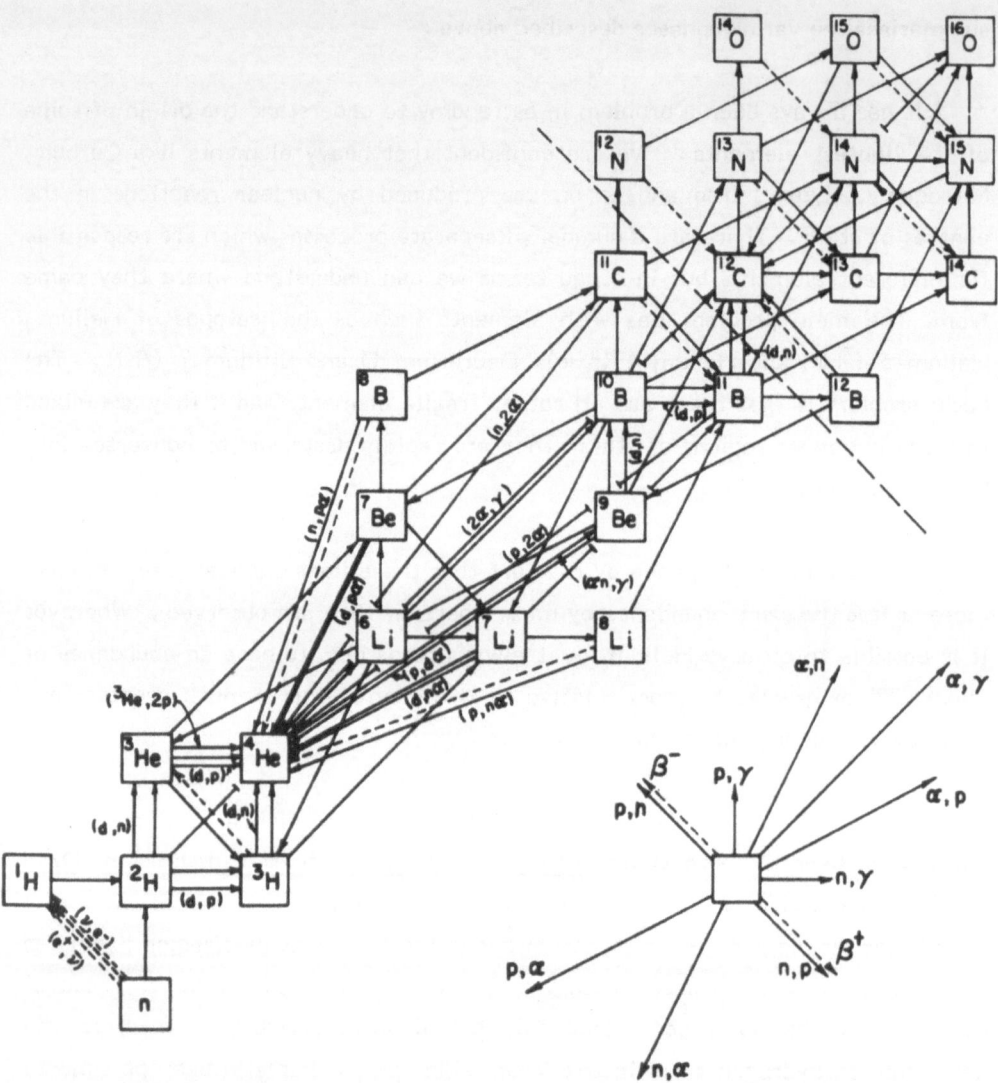

Figure 20   Illustrating the nuclear reaction chains which are included in modelling the synthesis of the light elements in the early Universe. The various interactions are indicated on the insert in the diagram (from R.V. Wagoner, 1973 , Astrophysical Journal , <u>179</u> , 343) .

evolution , no matter how one contrives the model . We will see in a moment how crucial this figure and its constancy are .

Recently , a similar analysis has been possible for Lithium-7 from studies of the spectra of very old halo stars in our own Galaxy . Despite variations in other element abundances , the ^{7}Li abundance seems to remain constant in those stars in which we have good reason to believe that the surface material has not been mixed into the inner regions of the star . These very beautiful observations by the French astronomers Spite and Spite show that the material from which these very old stars formed must have already contained about 1 part in 10,000 million by mass of ^{7}Li .

Let us now investigate in more detail what we expect to come out of the hot big-bang model . We can start the evolution of the model at a high temperature , $T \sim 10^{11}$K , at which we expect all the constituents of the Universe which are stable at that temperature to be in equilibrium . Then we can let all the constituents interact and see what elements are produced as the Universe expands and cools . Of course , this whole siulation is carried out in a computer and one must include in the calculation all the possible interactions between the various constituents of the Universe . Figure 20 shows the **"nuclear reaction chain"** which must be included in such a calculation . It includes all the routes by which all the isotopes of stable and unstable elements can be formed and destroyed as well as the spontaneous decay of unstable species . The probabilities of all these processes are now known in considerable detail from studies by particle accelerators .

These computations which have mostly been carried out by Dr Robert Wagoner of Stanford University involve large amounts of computer time to follow the evolution of each species . An example of the time evolution of the chemical composition of the Universe is shown in Figure 21 . It is not appropriate to go into the details of why the elements evolve in the way they do except to note that there is a profound difference between element building in stars and element building in the hot big-bang . In stars , the synthesis of the elements takes place over millions of years and there is time for equilibrium to be attained among the

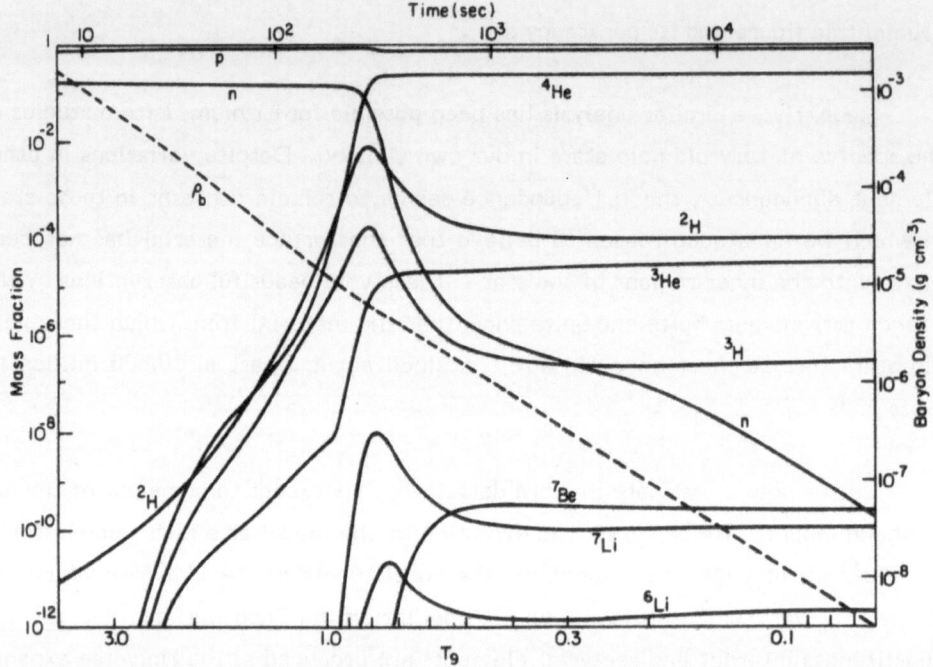

Figure 21   An example of the time and temperature evolution of the abundances of different species in the early evolution of the hot model of the Universe. Notice that the synthesis of elements such as D($=^2$H), ^{3}He, ^{4}He, ^{4}Li and ^{7}Be is complete after about 15 minutes (from R.V. Wagoner, 1973, Astrophysical Journal, <u>179</u>, 343).

different chemical species. In the hot big-bang, everything is highly non-equilibrium so far as the elements are concerned. The whole of the process of element formation is over in about 15 minutes as can be seen from Figure 21. This is because, by this time, the temperature has fallen below the value at which nuclear interactions can take place. In fact, the time interval in which element synthesis can take place is rather limited because at high temperatures there are so many hard gamma-rays that Deuterium, which is an essential first step in the formation of the light elements , is destroyed very rapidly . As the temperature decreases , more and more Deuterium survives and so can go on to form other light elements .

It turns out that the calculations only depend upon the ratio of the number of photons (or particles of the thermal background radiation) to protons (or baryons) in the Universe . Now , we know rather accurately the number density of photons in the microwave background radiation and so the results depend only on the present density matter in the Universe . The results are shown in Figure 22 .

The first remarkable feature of the diagram is that the elements which are synthesised in the hot big-bang are <u>exactly</u> those which are difficult to account for by stellar nucleosynthesis - i.e . D , ^{4}He , ^{3}He , ^{7}Li . Let us look at some other features . The ^{4}He abundance is remarkably independent of the density of the Universe and there are good thermodynamic reasons for this . For all reasonable values of the mean density of matter in the Universe , about 25% of ^{4}He is produced , in excellent agreement with observation .   Notice , in particular , the strong dependence of the Deuterium abundance on density . If the matter density is high , the Deuterium is all converted into ^{4}He and so , to obtain the present Deuterium abundance , a low value of the mean density of the Universe at the present time is required . Even more important is the fact that we know only ways of destroying Deuterium astrophysically and not of creating it . Therefore , if some of the Deuterium has already been destroyed , its original value must have been greater which drives one to lower values of the mean density of the Universe . On this basis , the density parameter for the baryons (or ordinary matter) must be less than $\Omega = 0.1$ i.e. <u>an open Universe</u> .

We can in fact select a best value for the baryon density of the Universe . I have indicated estimates of the abundances of the light elements on Figure 22 and it can be seen that a value $\Omega \sim 0.03$ can explain the Universal abundances of <u>all</u> the light elements . I find this a quite amazing result . It seems to me impossible that this could be a result of chance because the computations involve so many different interactions . I interpret these results as <u>independent</u> evidence that the

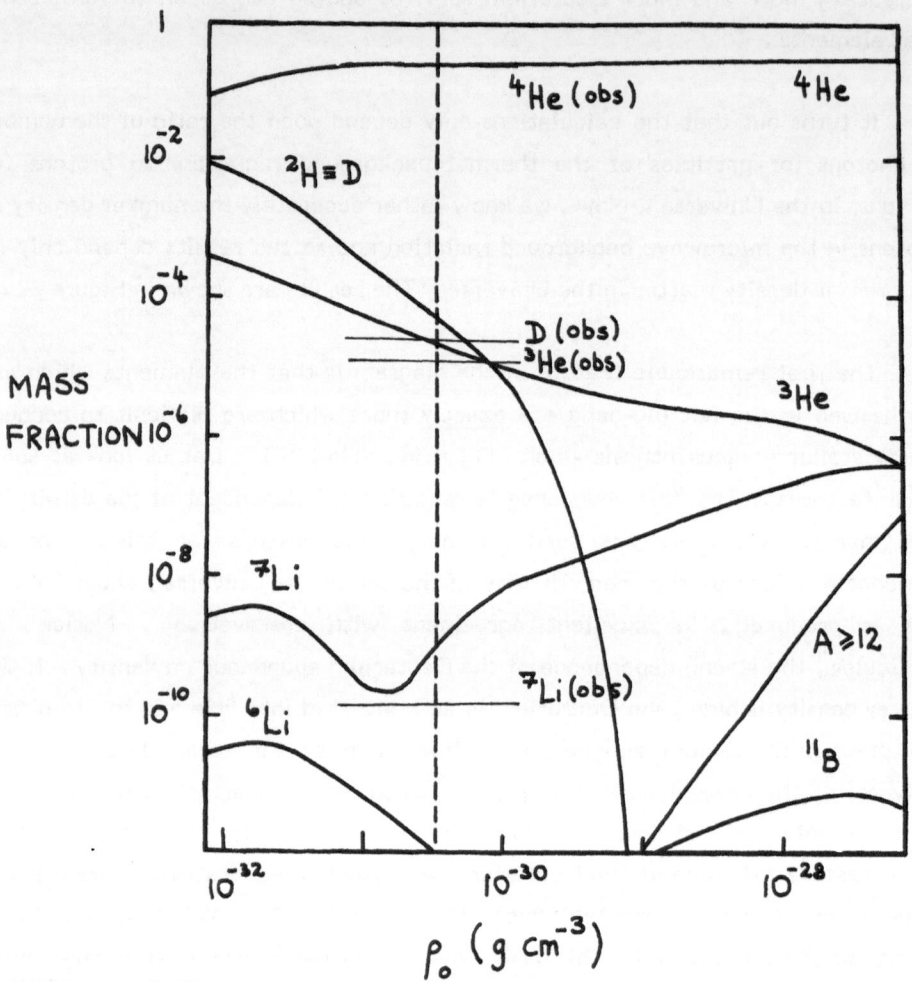

Figure 22   The predicted primordial abundances of the light elements compared with the observed abundances . The density of the world model at the present time is plotted along the abscissa .   The observed abundances are in good agreement with a model having $\rho_0 \sim 3 \times 10^{-31} \mathrm{g\ cm^{-3}}$ corresponding to $\Omega \sim 0.04$ (after R.V. Wagoner (op.cit.) and J. Audouze , 1982 "Astrophysical Cosmology" , eds . H.A. Brück , G.V. Coyne and M.S. Longair) , 395 , Pontificia Academia Scientiarum) .

Universe went through a very hot dense phase as described by the standard picture of the hot big-bang .

In this argument , the Deuterium abundance plays a critical role and this must be regarded as one of the great achievements of space science for cosmology .

The ramifications of this type of analysis for other aspects of cosmology and elementary particle physics are very wide but we will not go into them here . We note only one aspect and that is that the mean density of the Universe which comes out of these calculations may be lower than that inferred from dynamical estimates of Ω . The position is not yet clear but we should note that even if these estimates turn out to be different , this does not necessarily invalidate the above argument . This is because primordial element synthesis tells us only about the mass of baryons (or ordinary matter) and it could well be that there is additional mass present in some other form which would not influence the predicted element abundances . Examples of these possibilities are , small mass black holes , massive neutrinos and other ultra weakly interacting particles .

3.3 Great Problem No. 1 – the Baryon Asymmetry of the Universe

Already from this analysis we can identify one of the fundamental problems of the hot big-bang model . We have noted that in the very early stages of the model , $T > 10^{13}$ K , $R < 10^{-13}$, heavy particles and their antiparticles (baryons and antibaryons) are created in interactions between the photons of the thermal background radiation . In the earlier equilibrium phases , there will be roughly as many baryons and antibaryons as photons of the thermal background . This leads to the following somewhat strange result . From the analysis of the origin of the light elements , we know fairly precisely the ratio of photons to baryons in the Universe at the present day . If $\Omega_b = 0.03$ $H_0 = 50$ km s^{-1} Mpc$^-$ and the present temperature of the background radiation is 2.9 K , the ratio of the numbers of photons (n_γ) to baryons (n_B) is

$$n_\gamma/n_B \sim 10^9$$

Now, we have good evidence that virtually all the material of the Universe must consist of matter rather than a mixture of matter and antimatter. If the Universe did consist of intermixed matter and antimatter, there is a large possibility that they will annihilate by the opposite process to baryon pair production

$$p + \bar{p} \rightarrow \gamma + \gamma$$

This would lead to a vastly greater gamma-ray background intensity than is allowed by observations of the gamma-ray background. Therefore we can be confident that the Universe does not contain equal amounts of intermixed matter and antimatter. Another way of looking at this question would be to ask, "Suppose the Universe had started off with equal amounts of matter and antimatter, how many of the baryons and antibaryons would survive to the present day?" The answer is

$$n_\gamma / n_{B,\bar{B}} \sim 10^{18}$$

In other words, so much annihilation would have taken place that the ratio of photons to baryons would be wrong by a factor of 1,000 million.

Now let us take our matter Universe back to epochs before $t \sim 10^{-5}$ seconds. For every pair of photons a baryon-antibaryon pair is created and therefore the ratio of baryons to antibaryons must have been

$$\frac{n_B}{n_{\bar{B}}} = \frac{1000,000,001}{1000,000,000}$$

When this Universe evolves to later epochs, the 1000,000,000 baryons annihilate with the 1000,000,000 antibaryons producing 2000,000,000 gamma-rays and leaving only one baryon. This process ensures the present correct ratio of photons to baryons of about 1000,000,000 to one. It is this one baryon out of 1000,000,000 initial baryons and antibaryons which becomes the material Universe as we know it.

The problem which we have identified is that in the very early Universe we must build in a small asymmetry between matter and antimatter at the level of about 1 part in 1,000 million in order to ensure the correct ratio of baryons to photons now . The origin of this <u>lack of perfect symmetry</u> in the Universe is an important cosmological question . We will find that advances in particle physics give us some hope of understanding this problem .

Notice again a crucial input from space science . The gamma-ray background radiation imposes a strong limit upon the amount of antimatter in the Universe and reinforces the argument about the basic baryon asymmetry of the Universe .

Then this is the case for the hot big-bang model of the Universe and the first of our basic problems . Let us now try to construct a Universe with **real** objects in it .

4 . The Origin of Galaxies

The hot big-bang model of the Universe is still very dull because it contains no real objects like stars , galaxies , clusters , etc . We must now refine the model to explain the origin of galaxies .

Now , at this stage , the cosmologist washes his hands of most of the obvious things we have to make . In principle , we understand many of the basic things which must be going on in the internal workings of galaxies . We believe we understand the general picture for the evolution of stars from their birth to death. Figure 23 shows a schematic "time-lapse" photograph of the evolution of a star several times the mass of the Sun . The interval between frames is about one million years . Stars are born within dark dense dust clouds embedded within giant molecular clouds . Once they settle down to burning their nuclear fuel , which is the nuclear transformation of hydrogen into helium , they become what are known as **"main sequence" stars** and spend most·of their lifetime in this phase of evolution . Figure 23 shows this stage lasting about 100 million years . The

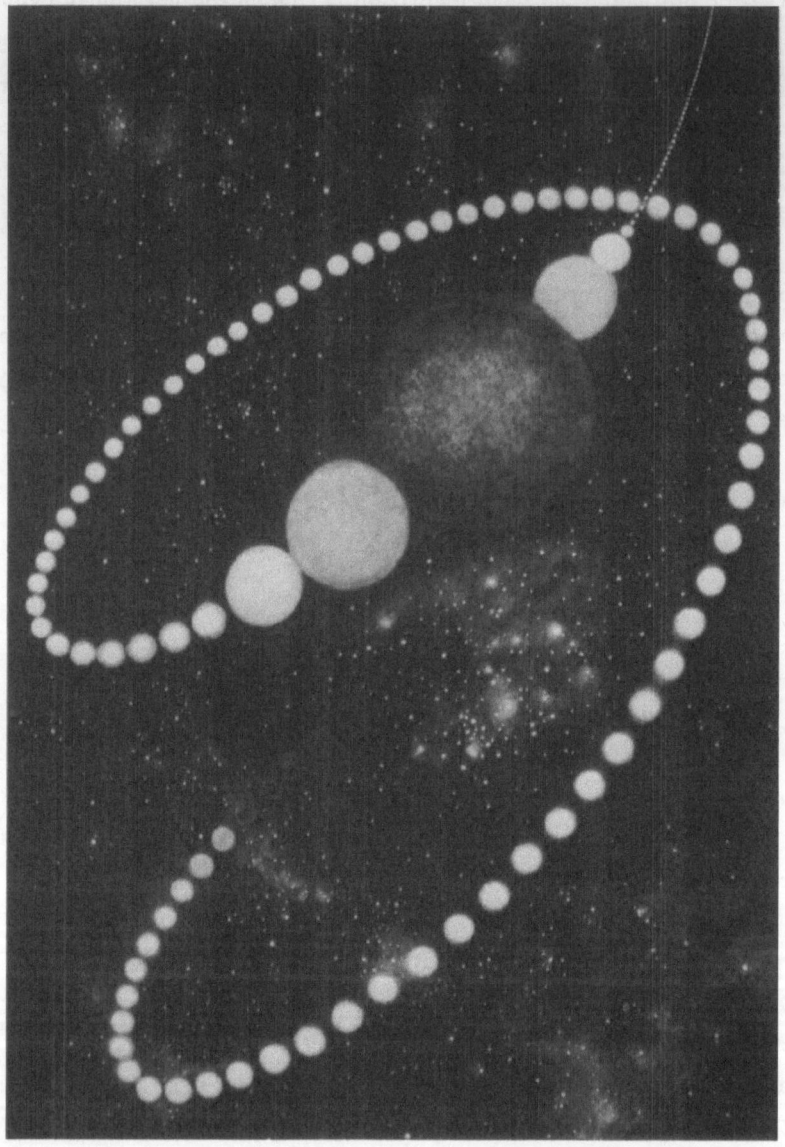

Figure 23   A "time-lapse" sequence of pictures of the evolution of a star several times the mass of the Sun .   The interval between images is about 1 million years (from H. Friedman , 1975 , "The Amazing Universe" , 76 , National Geographic Society , Washington) .

Sun is currently in this long hydrogen burning phase and is about half-way through its 10,000 million year lifetime as a main sequence star . When about 15% of the hydrogen of the star is burned in its core , the star undergoes an instability whereby the core contracts and the envelope expands . The large cool envelope of the star is very luminous and stars in this phase of evolution are known as **red giant stars** . The subsequent stages of evolution are rapid . There are phases when the star ejects mass from its surface which cause changes in its internal structure. However , whatever the details of the star's evolutionary history , it must eventually attain a lower energy state by collapse of its core . The end product of the final collapse can be one of three varieties of dead star - a **white dwarf** , a **neutron star** or a **black hole** . Astrophysically , an important part of the final stages of evolution is the ejection of mass from the evolving star to the interstellar medium . By this means , processed material is returned to the interstellar gas , enriched by the products of element building within the stars . The next generation of stars will form out of this enriched interstellar material .

We know that this cycle of birth , life and death must be repeated many times for stars in a galaxy such as our own and similar processes will take place in all types of galaxy . To the cosmologist , this means that he need not worry about making the detailed contents of galaxies . If he can create a large enough and dense enough gas cloud roughly the size of a galaxy to begin with , these astrophysical processes will take over and , we hope , create galaxies as we know them . It should be emphasised that there are many fundamental questions which must be solved before this picture can be made completely convincing in quantitative terms . Star formation is very poorly understood ; exactly which end points of stellar evolution are formed from different types of star is not clear ; the interchange of matter between stars and the interstellar medium is not properly understood ; we are unclear about the role of infall of material into galaxies from the intergalactic medium .

To quantify what the cosmologist is trying to achieve , he wants to find a means by which <u>perturbations</u> in the matter content of the Universe can grow to a stage where they can collapse and become <u>isolated bound systems</u> . By a bound system , we mean one which forms a distinct system and which holds itself together by its own gravitational forces . To put it simply , if you can find a means by which density enhancements can be generated in which the density is twice the average density , the job is completed .

Now , this might seem a rather unremarkable objective but in fact it is fraught with profound theoretical difficulties . What we would like is what is known as an exponential instability . This is the type of instability one finds in a nuclear explosion where , in a simple case , one neutron initiates the fission of a heavy nucleus which releases two neutrons . Each of these neutrons can initiate fissions resulting in 4 neutrons , and so on . The instability is such that a very small seed can rapidly result in a growing number of particles . It is the energy released in this exponentially increasing rate of fission which causes an atomic bomb explosion . In the same way , we would like to be able to start with an infinitesimally small density perturbation which would then grow exponentially to large amplitude . Mathematically , we would write the increase in density $\delta\rho$ as

$$\delta\rho \ \propto \ e^{\alpha t}$$

where α is a positive quantity . The hope is that essentially random fluctuations are sufficient to initiate the growth of an instability which can grow to a finite size .

Now , **gravity** possesses the inherently unstable property of being a long range attractive force . If the region is big enough in size , self gravity will make the region unstable . The problem is that in an expanding matter-dominated Universe the perturbations of density do not grow rapidly enough - the growth is not exponential but only algebraic with time . In the case of the world model with critical density , $\Omega = 1$, the perturbations only grow as

$$\frac{\delta\rho}{\rho} \propto t^{2/3}$$

We can obtain some understanding of how this comes about if we return to the simple model we constructed for the whole Universe (Figure 14 ; Section 3.1) . We can apply exactly the same dynamical argument to a spherical bit of Universe

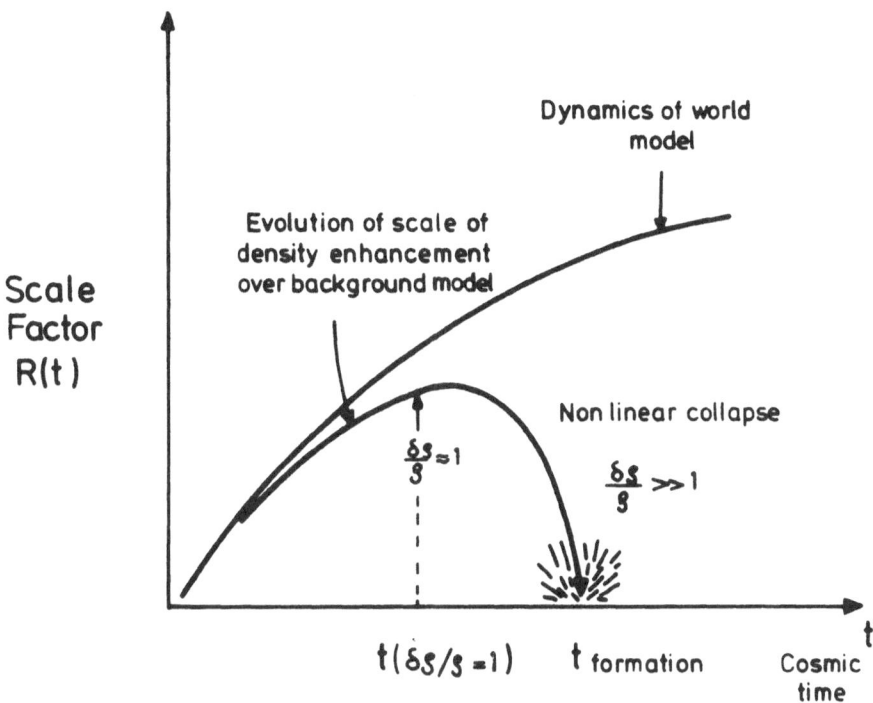

Figure 24   Illustrating the growth of a density fluctuation in a classical world model . The region behaves like a closed world model until the fluctuation grows to such a size that it can no longer be considered a small fluctuation .

---

with slightly higher density than the background medium . The spherical region will collapse more rapidly than the background medium as illustrated in Figure 24 . It is easy to show that the difference between the densities of the spherical region and the background does not increase exponentially but only algebraically , basically because the dynamics of the world models are described by algebraic functions . In the figure , we have shown the region collapsing non-linearly once the excess density is as large as the background density . By nonlinearity , we mean that the perturbation becomes so large that an analysis for small perturbations is no longer applicable . This means that one cannot begin with infinitesmally small perturbations but has to start with perturbations of finite amplitude . The obvious question then is "Where did they come from ?" .

This is nasty enough but matters are even worse . First of all , we have to ask how large physically the perturbations have to be before they can begin collapsing . This problem was first solved by **Jeans** and it results in the simple criterion that the size of the region should be such that the self gravitational forces in the region are greater than the pressure support forces within it . We can obtain a criterion for those scales which can become unstable by the following simple argument . Suppose we consider 1 cubic centimetre of material somewhere close to the edge of a region of size R and density ρ . Then the total mass of the region is roughly $M = \rho R^3$ and the force of gravitational attraction on our cubic centimetre of material towards the centre of the cloud is

$$F \; = \; \frac{GM\rho}{R^2} \; \sim \; GR\rho^2$$

Now , the forces which resist collapse are pressure forces , exactly the same sort of forces which maintain the Earth's atmosphere at atmospheric pressure . The pressure force acting on our centimetre cube of material is just the gradient of the pressure and to a rough (but good) approximation this is the pressure in the cloud   p   divided by the size of the cloud   R .

i.e . $F = p/R$.

We now obtain the criterion for collapse of the cloud . The gravitational force on the little cube of material should exceed the pressure support forces .

i.e . $GR\rho^2 \; > \; p/R$

or , reordering things

$$R^2 \; > \; \frac{(p/\rho)}{G\rho}$$

To complete the sum , we notice that to a good approximation p / ρ is just the square of the speed of sound in the cloud $c_s^2 \sim p / \rho$ and therefore we have a neat relation which tells us how big a region has to be before it will collapse under its own self-gravitation

$$R > \frac{c_s}{\sqrt{G\rho}}$$

This size is called the <u>Jeans' length</u> and the mass within the region is called the <u>Jeans' mass</u> .

   We can now work out which masses can collapse at various stages in the evolution of the Universe . At cosmological epochs after recombination (i.e . $z < 1500$, $R > 1/1500$), the Universe is matter dominated and we know the temperature of the gas is about 4000 K at that time . Putting the appropriate values into the above relation , we find that all masses greater than about 1 million times the mass of the Sun can collapse . This is an encouraging result . It means that regions the size of galaxies ($\sim 10^8 - 10^{12}M_\odot$), clusters ($\sim 10^{13} - 10^{15}M_\odot$) and even the larger structures can form by gravitational instability after the epoch of recombination .

   However , before the recombination epoch , the picture is very different . It will be recalled that there is strong thermal contact between the matter and radiation content of the universe prior to the recombination epoch and that almost simultaneously the Universe becomes radiation dominated . The matter becomes unimportant dynamically and the internal pressure in any cloud is dominated by the pressure and density of the radiation . In fact , the relevant sound speed to include in the above relation is the sped of sound in a gas consisting wholly of radiation which is roughly the speed of light , c . This means that collapse only occurs on the very largest scales $R_J > c/\sqrt{G\rho}$ where the radiation itself contributes the density ρ . Let us reorganise the relation a bit further to see what it means . We recall that for radiation $\rho \propto R^{-4}$ and in the radiation dominated phase $R \propto t^{\frac{1}{2}}$. Therefore , we find that the size of region $R_J \propto t$. In fact it turns out to be just less than ct where c is the velocity of

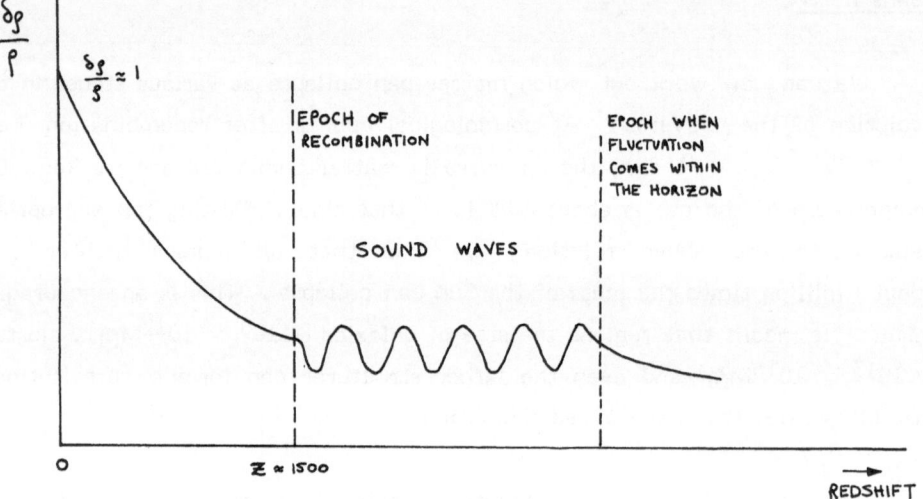

Figure 25   The evolution of adiabatic fluctuations in the expanding Universe. Prior to recombination , they behave as sound waves ; after recombination , the density enhancement grows as $t^{2/3}$ until redshifts $z \sim \Omega^{-1}$.

light . This is the origin of the profound problem about the origin of structure in the Universe as we now demonstrate .

The quantity ct is just the **"size"** of the Universe at epoch t . It tells us how far a light wave can travel in the universe at that epoch . It is known as the **"horizon distance"** in that it is as far as one can possibly communicate at epoch t . By communicate , I mean not only transmit information but transmit all physical forces in the Universe . The velocity of light is the maximum velocity at which this information can travel . We will come back to profound problems associated with this concept in Section 7 .

The problem which arises is that in the expanding Universe , the horizon size expands just as ct and so includes more and more matter in the observable Universe as time increases . However , according to Jeans' criterion in the radiation dominated epochs , almost as soon as a particular scale comes within the horizon , the size of the region becomes less than R_J which is almost ct . Since scales less than R_J are stable , the perturbation is so stabilised by the "springyness" of the gas of photons . The upshot of all this is that prior to recombination , perturbations on all scales are stable and do not collapse .

The problem of forming galaxies and clusters now becomes apparent . Just as an example , suppose we form galaxies at the present epoch , $R = 1$, $z = 0$. We will assume this simply means $\delta\rho/\rho \sim 1$. Now we know that the perturbations grow in the post-recombination epoch as R i.e. when $R = 1/1500$, $\delta\rho/\rho \sim 1/1500$. However , prior to that epoch , the fluctuation does not grow at all . It is simply a sound wave of almost constant amplitude. This means that to make anything in the Universe at the present day , we have to put in fluctuations with amplitude about $1/1000$ or $1/10,000$ in the very early Universe. This behaviour is shown schematically in Figure 25 .

This conclusion is very unpleasant for the cosmologist . It may seem as though these fluctuations are quite small , but they are absolutely enormous compared with the statistical fluctuations we would expect in any gas of particles. According to the classical picture , we have to endow the initial conditions of the Universe with further structure . In other words , give the Universe some more specific properties in its earliest stages . This is **Great Problem No. 2** ; the origin of the initial fluctuation spectrum which will eventually become the Universe as we know it .

Despite this major problem , we can ask how we expect these perturbations to evolve in the Universe and what the observable consequences of this evolution might be . It turns out that there are basically two alternatives depending upon the nature of the perturbations which are put in at the beginning . It is rather pleasant that these two alternatives correspond to quite distinct pictures for the origin of galaxies .

In one case , the fluctuations are what are called "adiabatic" fluctuations and are essentially just sound waves in the radiation-dominated plasma at pre-recombination epochs . It turns out that all the small scale sound waves are damped out by the epoch of recombination so that only sound waves on the very largest scales corresponding to clusters of galaxies and greater can survive into the post-recombination era . At this time , the internal pressure of the radiation suddenly declines because the radiation and the matter are no longer strongly coupled and the large scale perturbation begins to grow . They grow to large amplitude $\delta\rho/\rho \sim 1$ at a much later epoch . In the subsequent non-linear stages of collapse , the cloud fragments into smaller regions which will become galaxies . Thus the basic feature of this picture is the late formation of the largest systems which eventually fragment into the objects we know today .

The other case is that of "isothermal" fluctuations in which the seeds are not sound waves but simply regions of enhanced matter density . In this case , there is no damping of their amplitudes and fluctuations of mass about 1 million solar masses can survive to the recombination epoch . We have shown above that after recombination all masses greater than about 1 million solar masses can collapse after this time and consequently the smallest masses will collapse first . In this picture , the first things which form are masses of the order of star clusters rather than galaxies and they can form at large redshifts $Z \sim 100\text{-}1000$. These systems then cluster to form galaxies and larger scale

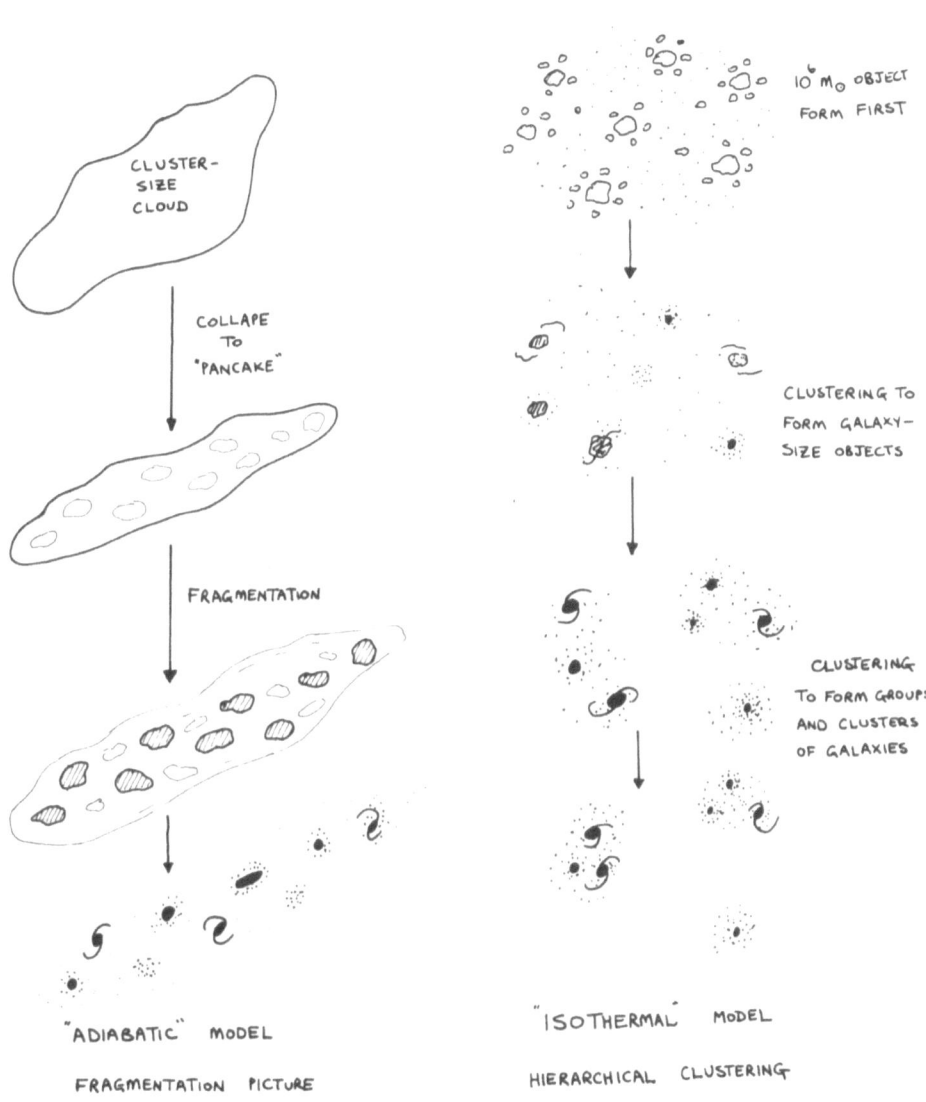

Figure 26   Illustrating two contrasting pictures of the origin of the large scale structure of the Universe .

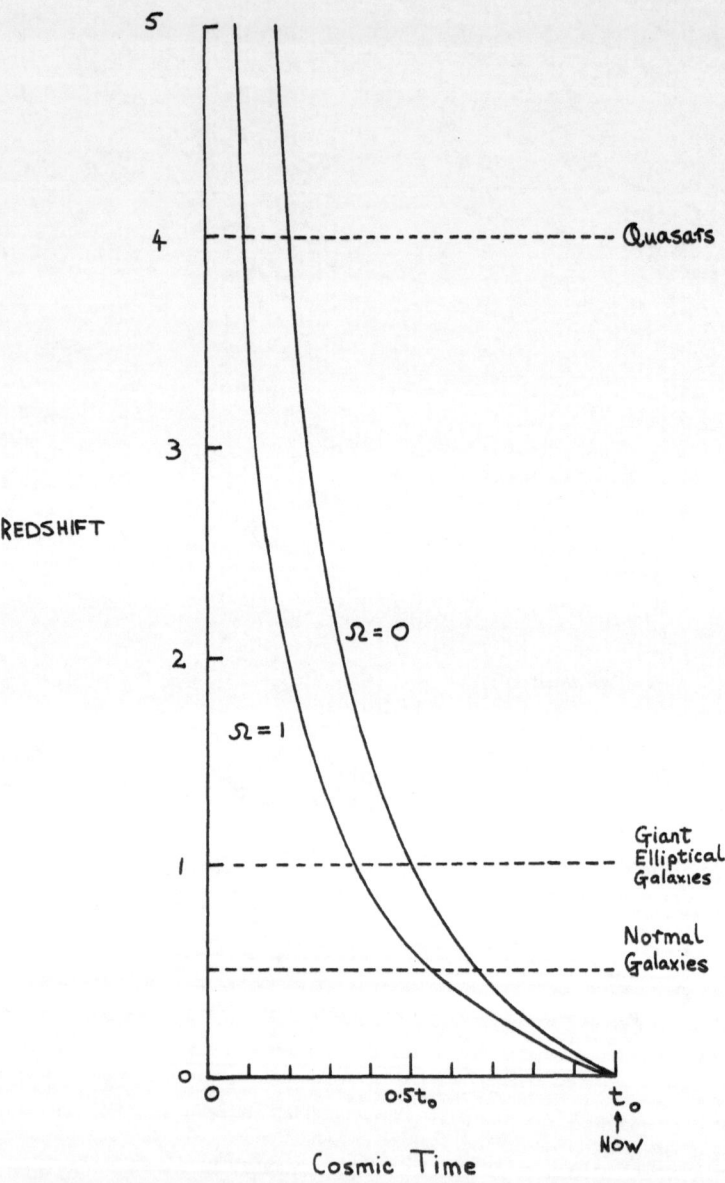

Figure 27   The relation between cosmic epoch and redshift for two representative world models . The redshifts to which different classes of object can be observed are indicated .

structures by the process of hierarchical clustering. Galaxies themselves form early in this picture.

These two different scenarios are shown schematically in Figure 26. An important goal for observational astronomers and cosmologists is to find out from observation which, if either, of these pictures is the best description of the sequence of events which led to the structure of the Universe which we observe today.

5. Evolution of Astronomical Objects in the Universe Over Cosmological Time-Scales

These theories of the origin of structure in the Universe provide additional stimulus for observers to probe to the faintest objects detectable in order to look back as far into the past as possible. It is interesting to note how far back in time we can look in the Universe. As we have emphasised this is a much more meaningful way of thinking about objects with redshifts of one and greater. The only problem with this approach is that the time corresponding to a particular redshift depends upon which world model we select. For reference we show in Figure 22 the results for two typical world models, one with $\Omega = 1$, i.e. the critical model which just expands to infinity and the empty model $\Omega = 0$ which is a good approximation for low density world models at small redshifts, specifically redshifts $z < 1/\Omega$.

The relations shown in Figure 27 also show the redshifts to which we can observe various types of object in the Universe. We can observe ordinary galaxies like our own out to redshifts of about 0.5 but then they become too faint to be observed by ground based telescopes. The most massive galaxies in the Universe such as those in the centres of rich clusters of galaxies and those associated with strong radio sources can be observed to redshifts greater than 1 but then they too become too faint. Notice that at a redshift of 1, Figure 27 tells us that these galaxies emitted the light which we observe now, when the Universe was either half its present age, if $\Omega = 0$, or when it was only about

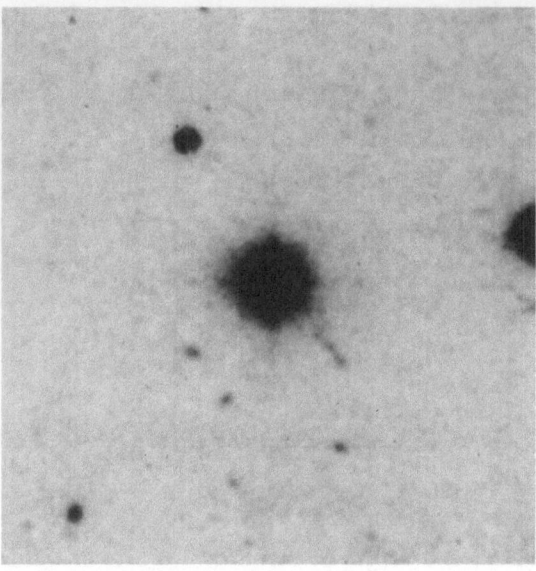

Figure 28   The quasar 3C 273 .   A faint jet can be seen extending from the quasar to the bottom right of the picture .

---

one third of its present age if $\Omega = 1$. Thus by studying the most luminous galaxies , we are looking at the Universe as it was when it was significantly younger than it is now and we might expect the properties of these galaxies to have changed significantly over this time scale .

To investigate the Universe when it was even younger , we need yet brighter objects and fortunately the **quasars** fulfil precisely this role . Quasars were discovered in the early 1960s and are "star-like" objects which are indistinguishable from ordinary stars on a photographic plate . 3C 273 , the brightest of all the quasars , is shown in Figure 28 . However , they are actually very distant objects. The reason they are so bright and star-like is that their optical emission is not ordinary starlight but results from the radiation of very energetic particles in active galactic nuclei . We know of a number of galaxies in which a great deal of similar activity is going on in their nuclear regions . One may therefore think of the quasars as being those systems in which the nuclei are hyperactive in the sense that the total optical emission from the compact central regions far exceeds the total starlight from a giant elliptical galaxy by a factor of 100 or 1,000 . These are the most luminous known objects in the Universe and they can be readily observed at large redshifts . The most distant known quasar in the Universe was discovered earlier this year by Ann Savage , Bruce Peterson and their colleagues and is known as PKS 2000-330 ; it has redshift z = 3.78 . For simplicity , let us work out the time when a quasar redshift 4 emitted its light . If Ω = 0 , the quasar emitted its light when the Universe was only one fifth of its present age ; if Ω = 1 , the Universe was less than a tenth its present age (see Figure 27) . Thus , there is no question but that when we study very distant quasars we are investigating objects at very much earlier stages in the evolution of the Universe than the present epoch .

It has been known for some time now that the quasars and radio galaxies show strong cosmological evolutionary changes with cosmic epoch . The first evidence came from studies of extragalactic radio sources , the same types of object which dominate the picture of the radio sky . The brightest of these radio sources are already at cosmological distances as the histogram of Figure 29 shows . Among the brightest 100 radio sources in the northern sky , roughly half of them are at redshifts greater than 0.5 and so by looking at fainter samples of sources , it becomes possible to study the same objects but at much greater redshifts . These studies , known technically as radio source counts , showed that the violent events which gave rise to radio galaxies and quasars were much more common at early cosmological epochs than at the present time .

The quasars show exactly the same effects . The evidence for these evolutionary changes is shown in Figure 30 where a test known as the V/V_{max} test is carried out for a sample of the brightest radio galaxies and quasars . The V/V_{max} test is an elegant test for the uniformity of the distribution of a set of objects in space . If the objects observed within a complete statistical survey are uniformly distributed , then we expect that on average we should see them halfway to the limiting volume within which they could have been seen . If the sources had been further away , they would have been too faint to be included in the survey . In fact if we form the quantity V/V_{max} where V is the volume of

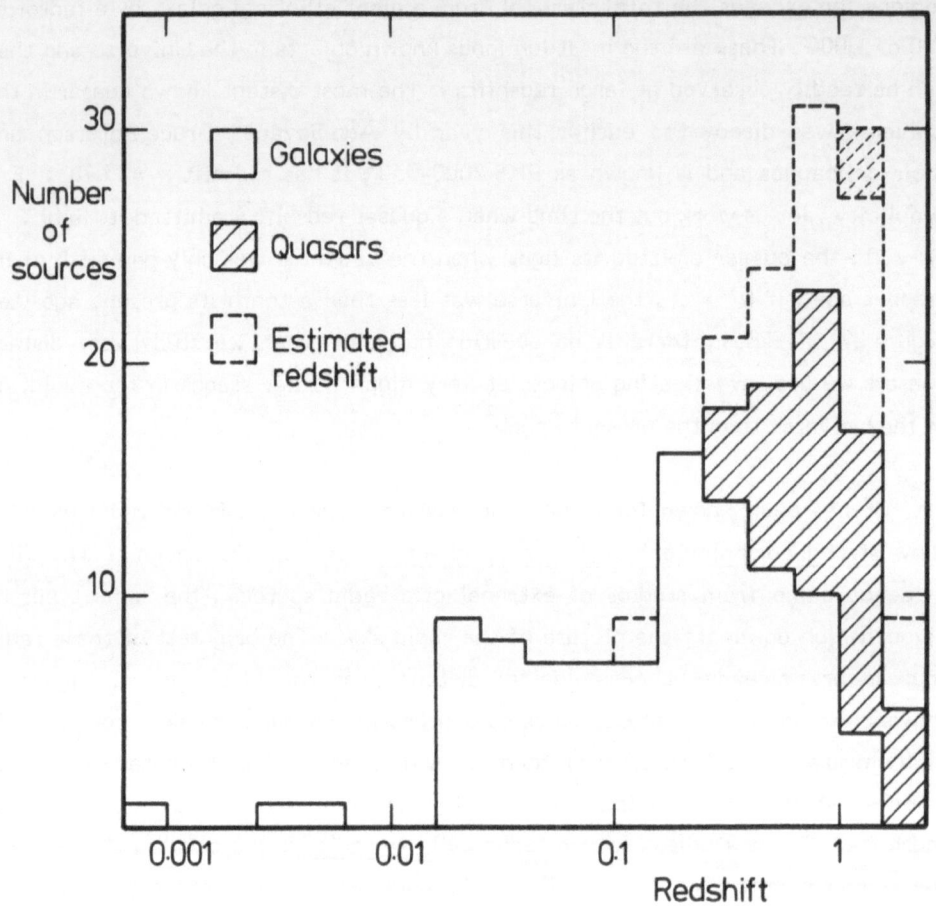

Figure 29   A histogram showing the redshift distribution of the 170 brightest radio sources in the sky in directions away from the Galactic plane . The radio galaxies and quasars are indicated by different symbols .

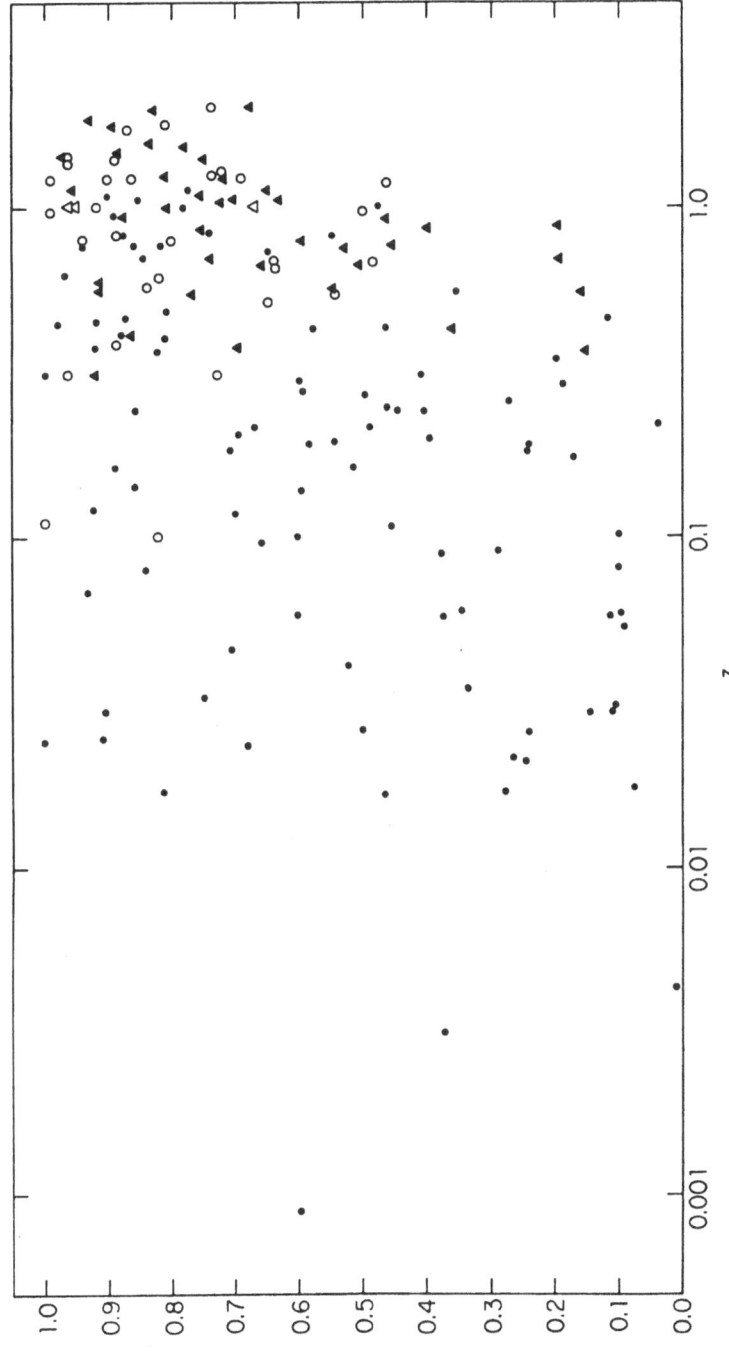

Figure 30   The V/V$_{max}$ test for the 170 bright radio galaxies and radio quasars shown in Figure 29 .   The galaxies are indicated by circles and the quasars by triangles (from R.A. Laing , J.M. Riley and M.S. Longair , 1982 , Monthly Notices of the Royal Astronomical Society) .

space lying within the distance of the source and V_{max} the maximum volume within which the source could have been observed and still remain within the sample , the values of V/V_{max} should be uniformly distributed between 0 and 1 if the sources are uniformly distributed in space . The actual distribution as a function of redshift for bright radio sources and quasars is shown in Figure 30 .

It can be seen that at redshifts less than 0.5 , the points are rather uniformly distributed and the average value of V/V_{max} is about 0.5 . However at all larger redshifts , the points show a strong tendency to have values greater than 0.5 , the average value being closer to 0.7 . This result applies to both the radio galaxies and quasars which are indicated by different symbols . This means that radio galaxies and quasars are piled up towards the limits of their observable volumes and since the volumes extend to cosmological distances , there must have been many more of them in the past .

Exactly the same analysis is possible for those quasars which are selected entirely according to their optical properties and they too exhibit a piling up towards the limits of their observable volumes .

It is not a straightforward matter to interpret all these data which indicate that there was much more violent activity in the past . In general , they imply quite enormous evolutionary changes in the average properties of these active systems with cosmic epoch .   The types of change necessary to explain these observations are shown in Figure 31 . In round terms , the probability of quasar activity being found in galaxies or of strong radio emission being generated was about 1,000 or 10,000 times greater when the Universe was about a quarter or a fifth of its present age than it is now . In other words , the active galaxies we see now are very much the tail-end of a period of very violent activity which occurred when the Universe was only about a quarter or a fifth of its present age . If we had been present in the Universe at this earlier time we would have seen thousands more quasars and radio galaxies than we see today .

There is one very intriguing aspect of Figure 31 . It shows that there is likely to be cut-off to this violent activity at a redshift of about 3-4 . It should be noted that this result comes from interpretations of the radio source counts and the V/V_{max} data for radio galaxies and quasars . It is remarkable that the data for optically selected quasars show exactly the same effect . There appears to be a deficit of quasars with redshifts greater than about 3.5 . For 10 years , the upper limit to the redshift of quasars was 3.53 and despite intensive optical searches designed specifically to find larger redshift quasars , none was found . It was only earlier this year that a larger redshift quasar was discovered and that was by the radio optical identification technique rather than by specific optical searches . A statistical analysis of the optically selected quasars alone suggests that there is a real deficit of large redshift quasars .

It is intriguing that these two independent methods of investigating the cut-off at large redshifts lead to similar positive conclusions . They are independent in the sense that they use completely different classes of objects - one of them is wholly an optical technique and the other is almost entirely dominated by the radio evidence . This tends to suggest that there is a real cut-off in the distribution of all types of active nuclei at about this redshift . It suggests that it is a real physical cut-off rather than being due to some form of obscuration because the optical and radio waves are affected in very different ways by obscuring matter .

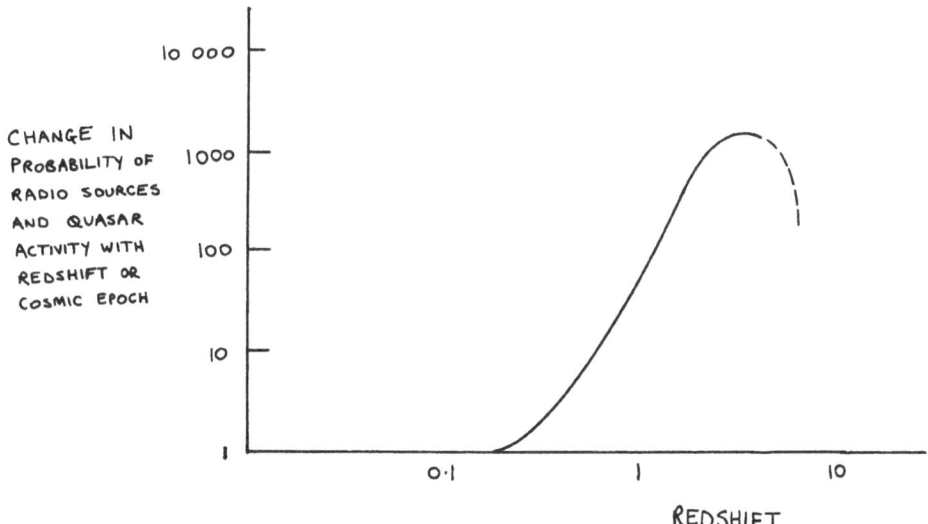

Figure 31   Illustrating the changes with cosmic epoch of the probability of there being strong radio sources in the Universe .

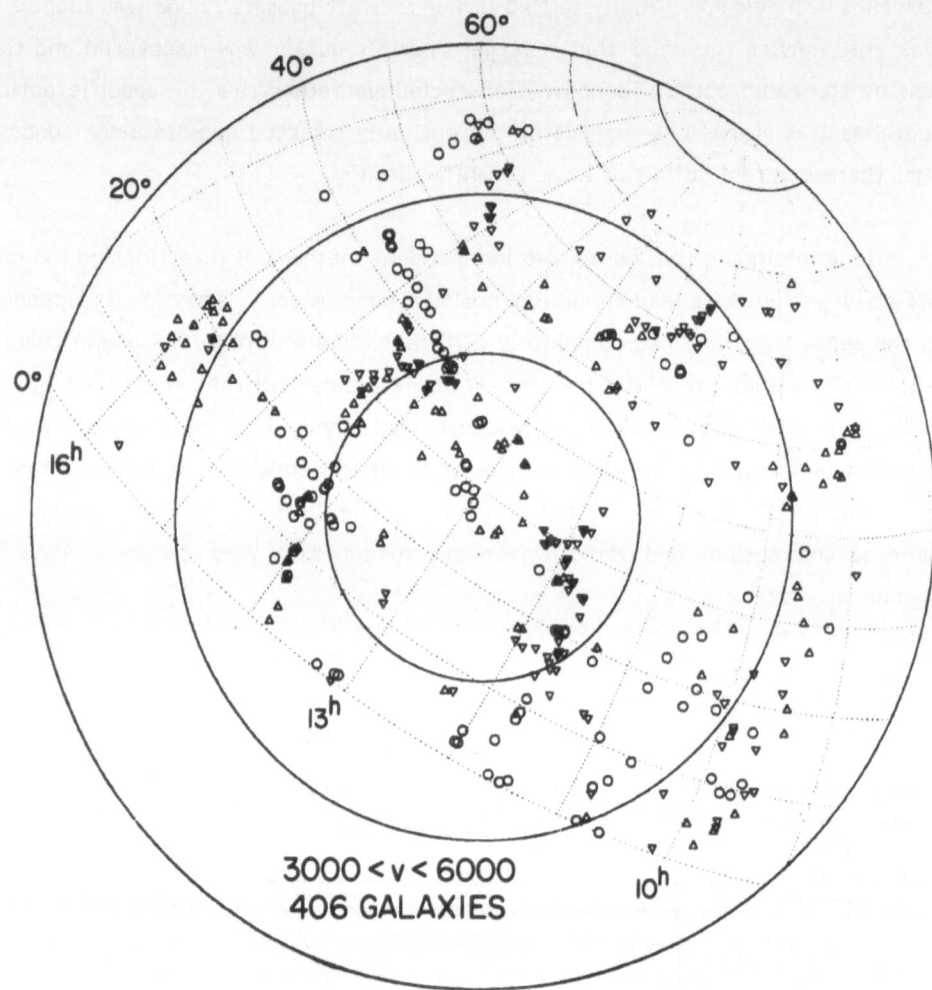

Figure 32   The distribution on the celestial sphere of galaxies with velocities in the range 3000 to 6000 km s^{-1} corresponding to redshifts roughly 0.01 to 0.02. The "filamentary" structure and large holes can be seen in the distribution of galaxies (from M. Davis, "Astrophysical Cosmology", (eds. H.A. Brück, G.V. Coyne and M.S. Longair), 117, Pontificia Academia Scientiarum).

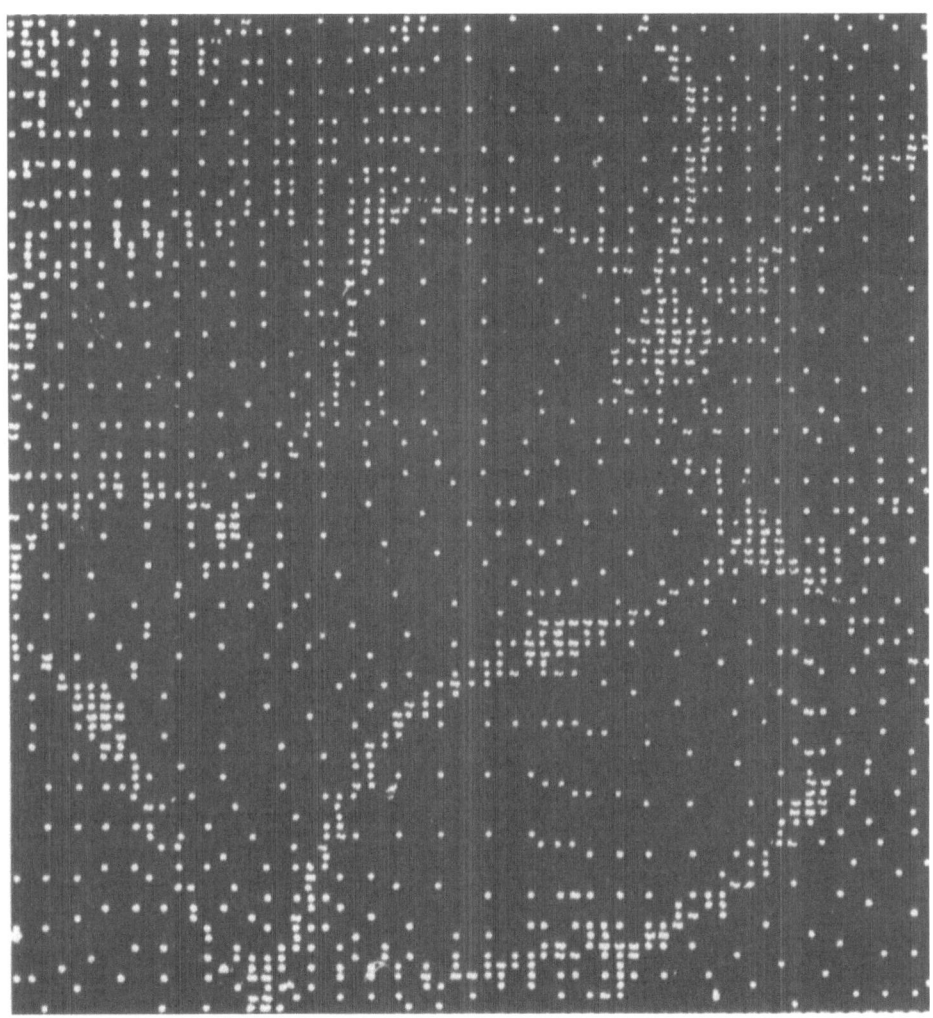

Figure 33   Simulation of the collapse of "pancakes" in the adiabatic picture of tne formation of large scale structures in the Universe (from Ya. B. Zeldovich , "The Large Scale Structure of the Universe" , IAU Symposium No. 79 , (eds . M.S. Longair and J. Einasto) , 405 , D. Reidel Publishing Company) .

A natural interpretation of this phenomenon is that it is related to the epoch when galaxies were first forming . My own belief is that one needs a major event of this magnitude to account for the sudden onset of activity of extreme violence in the Universe and then to account for its sudden decline . If this interpretation is correct , it has profound implications for observational and theoretical astronomy . First , it means that the processes of galaxy formation may well lie within reasonably accessible redshifts and hence the theory can be confronted with real observational evidence . Second , it favours models in which galaxies form late in the Universe which correspond to the fragmentation picture described in Figure 26 .

There is one other piece of evidence which can be naturally accounted for by this scenario . Zeldovich and his colleagues in Moscow have followed in some detail the evolution of the fragmentation (or adiabatic) scenario discussed above . There is one rather distinctive prediction of that theory which can naturally account for the filamentary structure of the Universe seen in Figure 7 . This observational result has been confirmed by surveys of the distances of bright galaxies which show that in space , as well as on the surface of the celestial sphere , the galaxies lie along **"chains"** and on **"cell-walls"** rather than being randomly distributed . An example of this is shown in Figure 32 . The origin of this phenomenon may be understood as follows . When one of the very large structures begins to collapse , it is most unlikely that it will be spherically symmetric. There will generally always be one shortest axis and Zeldovich has shown by a remarkable argument that collapse occurs preferentially along this shortest axis. This means that when perturbations collapse , they will tend to form sheets or **"pancakes"** rather than spherical systems . The galaxies then form preferentially by fragmentation within these sheets . If one can imagine this process taking place in three dimensions , it will lead to a three-dimensional filamentary structure similar to that observed in Figure 7 . A computer simulation of this process has been carried out by Zeldovich and his colleagues (Figure 33) and this shows distinct similarities to Figure 7 . It is not so easy to account for these structures in other theories of the origin of the large scale structure of the Universe .

My personal view is that both of these arguments favour the fragmentation picture in which galaxies form late in the Universe . It would be wrong to claim that other pictures can be excluded at this stage but at least we now have some real observations with which the theories can be confronted .

If this picture is correct , we might expect to see some effects of **evolution** of the properties of ordinary galaxies themselves at the redshifts which we can now observe with large telescopes . There is indeed accumulating evidence that this is observed from the few systematic surveys of large redshift galaxies which have been carried out . The main problem is finding large redshift galaxies in a systematic way since any given very faint galaxy may either be a very distant highly luminous galaxy or a nearby faint galaxy . One way of finding these distant galaxies is to use radio galaxies which we know to be among the most luminous of all galaxies . They are also among the very few of these distant galaxies for which redshifts have been measured . Our own work , by Simon Lilly and myself , has concentrated upon those radio galaxies in which we can be sure that the light of the galaxy is associated with its normal stellar population rather than being associated with a quasar-like component in the nucleus . We have studied this problem by comparing the colours of these galaxies in the infrared and optical wavebands . We know the spectra of nearby galaxies and can predict how their colours should change with redshift . The predictions and observations are shown in Figure 34 . If galaxies do not change their properties with redshift , they should follow the line labelled NE meaning "no evolution" . It can be seen that the galaxies lie along this line nicely at redshifts $z < 0.4$ but at large redshifts (i.e. at earlier cosmic times) the observations depart from this prediction. Instead , the galaxies are much "bluer" in the sense that there is more visual light than would be expected from the infrared luminosity . This evolution of the spectrum of radio galaxies can be modelled by supposing that the birth rate of stars was greater in the past . An example of these models is that labelled 0.5 in Figure 34 . In this case , there is an early burst of star formation which lasts 1,000 million years after which half of the mass of the galaxy is in the form of stars. The rate of star formation then declines exponentially to the present day . This behaviour is indicated schematically in the insert to Figure 34 . In this picture , when we look to a redshift of one , i.e. about half the present age of the Universe , the star formation rate was much greater than it is now .

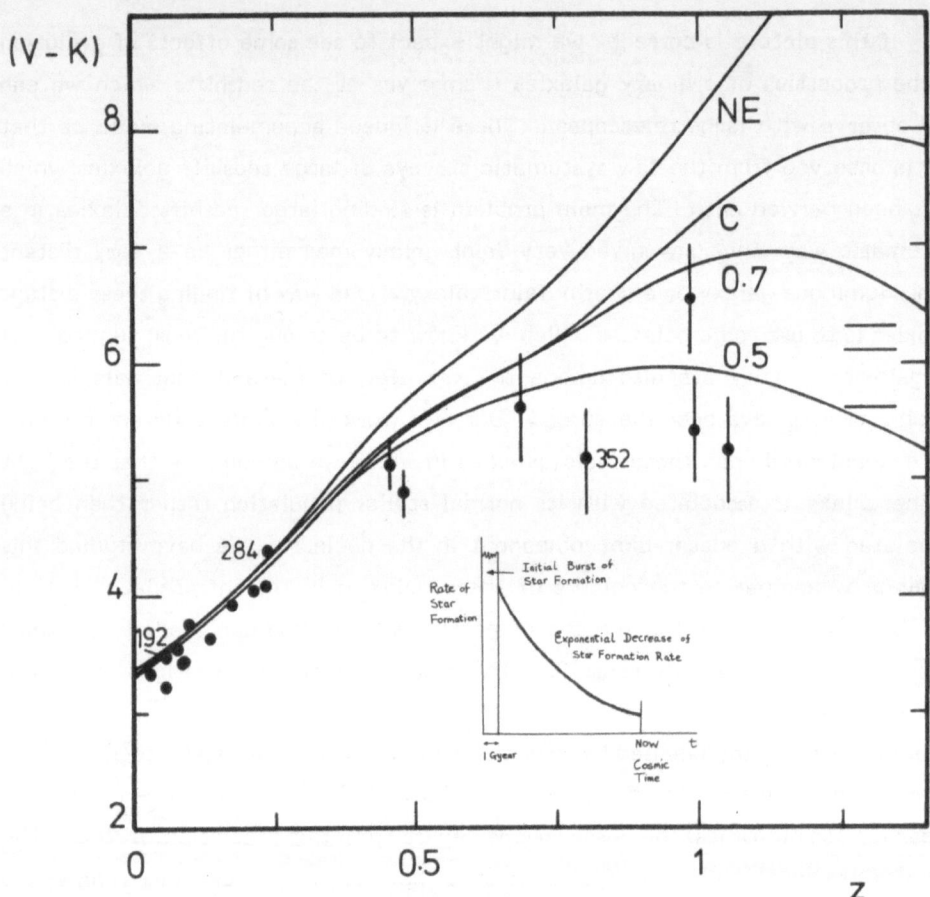

Figure 34   The variation of the optical-infrared colours of the stellar component of radio galaxies with redshift . The line labelled NE shows the expected colour variation if the galaxy's spectrum is unchanged with cosmic epoch . The model labelled 0.5 has an evolving rate of star formation which is illustrated in the insert (after S.J. Lilly , M.S. Longair and I.S. McLean , 1982 , Nature).

This is an intriguing result because it suggests that the ordinary properties of galaxies are changing rapidly with cosmic epoch. Indeed, it is remarkable that the variation of the rate of formation of radio sources can also be approximated by an exponential function over the same range of cosmic epochs and this is suggestive that there may be a causal connection between the two phenomena. Exactly how this came about is not clear because the radio sources originate in violent activity in the active nucleus, whereas the colour variations in the stellar component of the galaxy refer to ordinary stars and the amount of gas present in the galaxy as a whole. Nonetheless, the exciting thing is that at last, we may have some genuine observational clues to the types of phenomena which may contribute to the evolution of ordinary galaxies and to the violent activity originating in the nuclei of the most massive systems.

6. The Role of Space Observatories in the Study of Astrophysical Cosmology

We have made the case that Space Science has already made some vital contributions to cosmology. My own conviction is that many of the outstanding problems will best be addressed by missions in space, some of which are already underway and others which are at the approved or planning stage. In the light of the above discussion, let me take each waveband in turn and describe what I see to be the essential contributions to be made by space observations.

6.1 The Submillimetre Waveband

By this waveband, I mean the wavelength region from about 1 mm to 100 μm where the atmosphere is largely opaque for incoming radiation. Perhaps the most important of all experiments for cosmology are those which seek to establish with ultrahigh precision the spectrum and isotropy of the microwave background radiation. There are two essential issues : (a) Is the spectrum of the background radiation precisely a black-body spectrum or are there small deviations from it? (b) What evidence is there for anisotropies in the microwave background radiation on all angular scales ?

In the first case, the question of whether the spectrum is precisely Planckian is important in understanding whether or not there was ever a large input of energy into the Universe at epochs after about 100,000 years from the big-bang. This might prove to be one of the only ways of investigating energetic events which could have occurred in the Universe prior to the formation of galaxies as we know them.

In the second case, the anisotropy of the background is the subject of a wide range of different studies. The simplest case is what is known as the <u>dipole</u> component in which the sky is hotter in one direction and colder in the opposite direction. This phenomenon has now been convincingly observed from balloons and high flying aircraft and occurs at the level of about 1 part in 1,000. This is naturally attributable to the motion of the Earth with respect to the frame of reference in which the microwave background would be 100% isotropic. It is due to what is known as our "peculiar velocity" with respect to the frame of reference of a local fundamental observer i.e. one who would observe the background to be 100% isotropic. This velocity tells us something about the forces acting on our own Galaxy due to the large scale distribution of galaxies.

After removing this anisotropy due to our motion, the microwave background appears to be remarkably isotropic. However, the detection of higher moments of the microwave background radiation would be of profound cosmological significance since they are intrinsically bound up with the large scale topology of the Universe and the residue of any anisotropies in the very early Universe.

A third area of great importance is the study of anisotropies in the background radiation on small angular scales. It is predicted that these should be present at a low level because when large scale structures begin to collapse, they scatter the background radiation producing small temperature fluctuations along different lines of sight in the Universe. The detection of these fluctuations is a strong prediction of many theories of the origin of structure in the Universe. They have not yet been detected but I personally believe they should be observable with an order of magnitude increase in sensitivity. This experiment can only be successfully carried out in space.

Some of these questions will be addressed by the **NASA Cosmic Background Explorer (COBE)** which is scheduled for launch in 1987. It will survey the whole sky to look for anistropies on angular scales greater than 7° at the level of about one part in 100,000 rather than one part in 1,000 as at present. In addition, it will make very precise measurements of the spectrum of the radiation to an accuracy of about 1 part in 1,000. It is quite impossible to achieve anything like this precision from balloons or high-flying aircraft - it must be done properly from space. A clever experiment has been devised by Derek Martin and an international team of collaborators to measure the spectrum of the background with very high precision. This experiment known as the **Cryogenic Infrared Background Satellite (CIRBS)** still awaits approval but could be an excellent candidate for a Spacelab experiment.

The waveband from about 1mm to 100 μm is largely unexplored astronomically. With the advent of high sensitivity submillimetre telescopes on high altitude sites, this waveband will be opened up for astronomical observatories probably to wavelengths as short as 350μm. However, even on the best sites, the transparency of the atmosphere in the useable windows is at best about 50%. Systematic studies to shorter wavelengths require a space experiment and one currently under study is the **Large Deployable Reflector (LDR)** which would be a submillimetre telescope in space of more than 10m diameter (Figure 35). Four areas of study are particularly interesting for cosmology. First, many galaxies are now known to emit most of their energy at wavelengths of about 100μm. This is the result of the optical and ultraviolet radiation being absorbed by dust and reradiated in the far infrared and submillimetre wavebands. With a space submillimetre telescope such as the LDR, these galaxies would be observable at cosmological distances. Second, most active galaxies probably emit most of their radiation at about 100μm and hence to measure the total energy output of these objects in the Universe we need to know exactly how much energy is emitted by all classes of object in this waveband. Third, small deviations from perfect smoothness of the microwave background radiation may be due to the radiation scattered by hot gas clouds among the line of sight. The detection of these very small perturbations would provide a completely new tool for the study of very hot gas clouds at different stages in the evolution of the

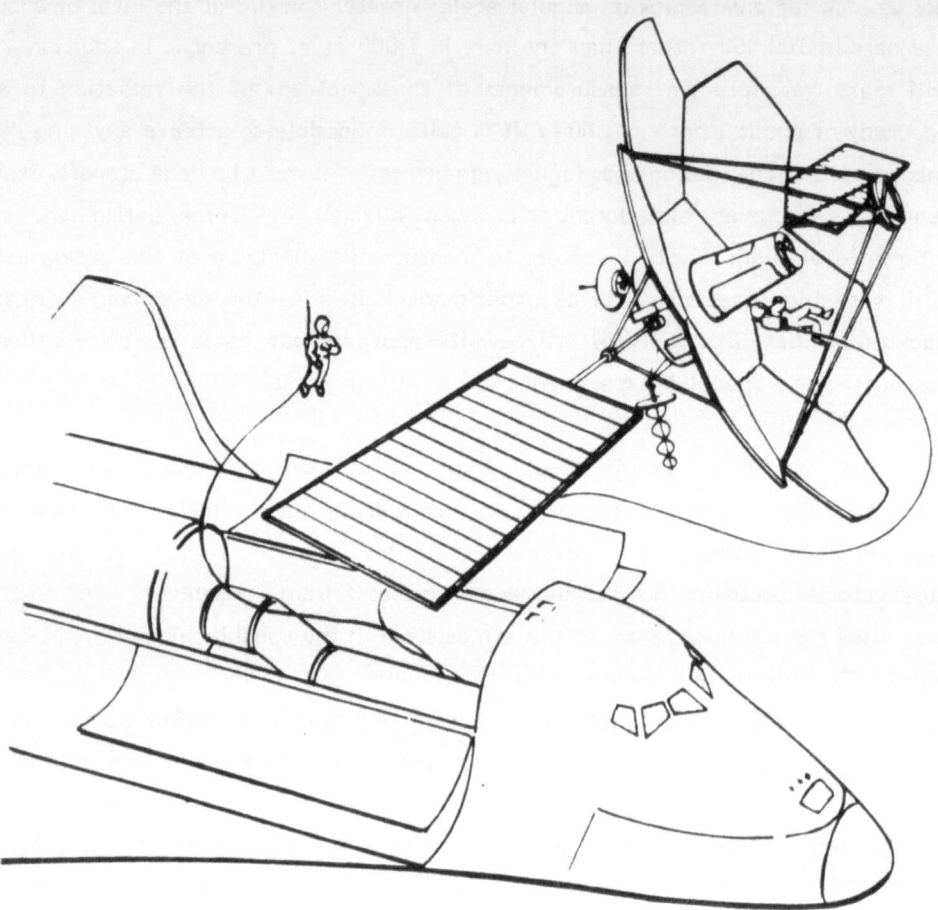

Figure 35   An artist's concept of the Large Deployable Reflector (from J.J. Russo and A.N. Bunner , 1981 , Perkin Elmer Technical News , 9 , 39) .

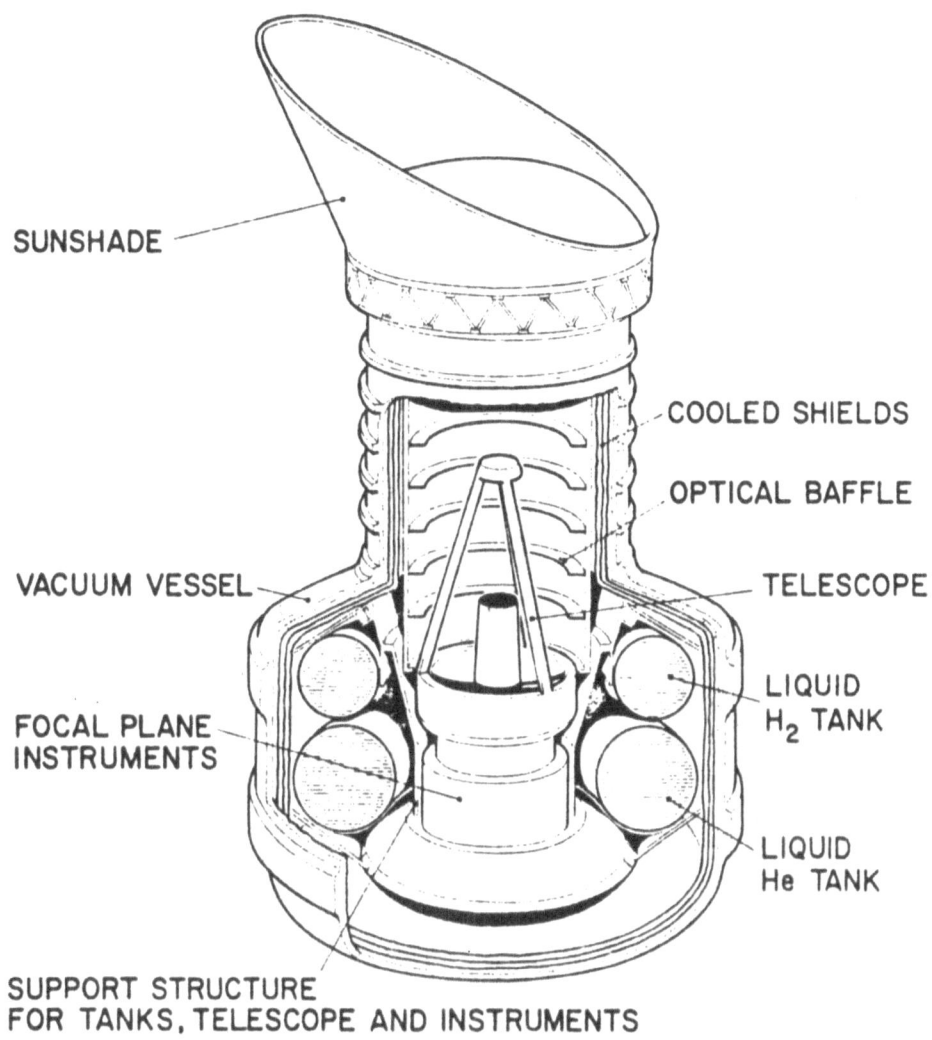

SUNSHADE

COOLED SHIELDS

OPTICAL BAFFLE

VACUUM VESSEL

TELESCOPE

LIQUID H$_2$ TANK

FOCAL PLANE INSTRUMENTS

LIQUID He TANK

SUPPORT STRUCTURE FOR TANKS, TELESCOPE AND INSTRUMENTS

Figure 36   A schematic diagram showing a concept for the ESA Infrared Satellite Observatory (from R. Emery , 1982 "The Scientific Significance of Sub-Millimetre Astronomy" , ESA publication , in press) .

Universe . Fourth , the LDR is exactly the type of telescope needed to search for fine scale anisotropies associated with the initial collapse of the large inhomogeneities soon after the epoch of recombination .

6.2 The Infrared Waveband   100μm to 1μm

There are several windows in this waveband within which successful observations can be made from the ground . It turns out to be remarkably straightforward to detect very distant galaxies in the 1-3μm waveband thanks to recent advances in infrared detector technology . However , in the thermal infrared wavebands (5μm to 100μm) , the observations are difficult because the atmosphere is very bright . The situation is similar to that of trying to see stars in the daytime . In both cases , it is the radiation (or reradiation) of the atmosphere which causes the problem and the way of achieving an immediate huge increase in sensitivity is to make observations from above the atmosphere .

The first complete satellite surveys of the infrared sky in the 5 to 100μm waveband will be made by the **Infrared Astronomy Satellite (IRAS)** which is due to be launched in early 1983 and which is a joint USA-Netherlands-UK project . The prime function of IRAS is to make a complete survey of the infrared sky . Of particular importance for cosmology will be the detection of thousands of galaxies in the infrared waveband for the first time . The most luminous of these may well be at cosmological distances .

For me , perhaps the most exciting challenges of the future are provided by the next generation of Infrared Satellite Observatories , for example the **NASA Cooled Infrared Telescope Facility (CIRTF)** and the ESA **Infrared Satellite Observatory (ISO)** (Figure 36) . These will be complete infrared satellite observatories with the capability of integrating for long periods on any region of the sky with high angular resolution . The gain over all previous missions , including IRAS which only integrates for a few seconds on each source with rather low angular resolution , will be enormous . They will have the capability of detecting most classes of extragalactic systems at cosmological distances .

One question which we have not properly addressed is the problem of detecting the first generations of galaxy which form in the Universe . One real possibility is that most of the energy is liberated in the infrared waveband , either because the sources are at very large redshifts or because the radiation is mostly reradiated dust emission in the thermal infrared wavebands . Despite searches in other wavebands , no convincing candidates for primordial galaxies have been discovered. Whilst I would not rule out the possibility that they will be discovered by the Space Telescope , I am somewhat surprised that there is not some evidence for them already . I believe there is a strong possibility that the thermal infrared waveband may provide the crucial clues we need to understand these problems and that satellites such as ISO are ideally matched to these studies .

6.3 Optical and Ultraviolet Wavebands

Up to the end of the present century and beyond , these wavebands will be dominated by the NASA **Space Telescope** (Figure 37) . This will be a 2.4m optical and ultraviolet telescope with diffraction limited angular resolution . In a word , this means that it will take pictures at least 10 times sharper than the best one normally achieves from the ground and also that , for the first time , deep images will be taken in the ultraviolet region of the spectrum from 1200 to 3000Å . The scientific significance of this project is enormous . In simple terms , it will enable astronomers to do the sort of astronomy they can now do from the ground on objects at least 10 times more distant than is possible now . Since we have already shown that we can observe objects at significantly earlier cosmological epochs , this capability raises the opportunity of studying systematically objects in the Universe at earlier epochs .

What makes this particularly exciting is the fact that at least some of the bright galaxies are showing evidence of cosmological evolution in the sense that they were more luminous and bluer in the past . Since we expect these effects to be even more important in the ultraviolet wavebands , we have every reason to expect that the deep ultraviolet images will contain many more bright galaxies at large redshifts than we would expect by a simple extrapolation from the properties of galaxies at the present day .

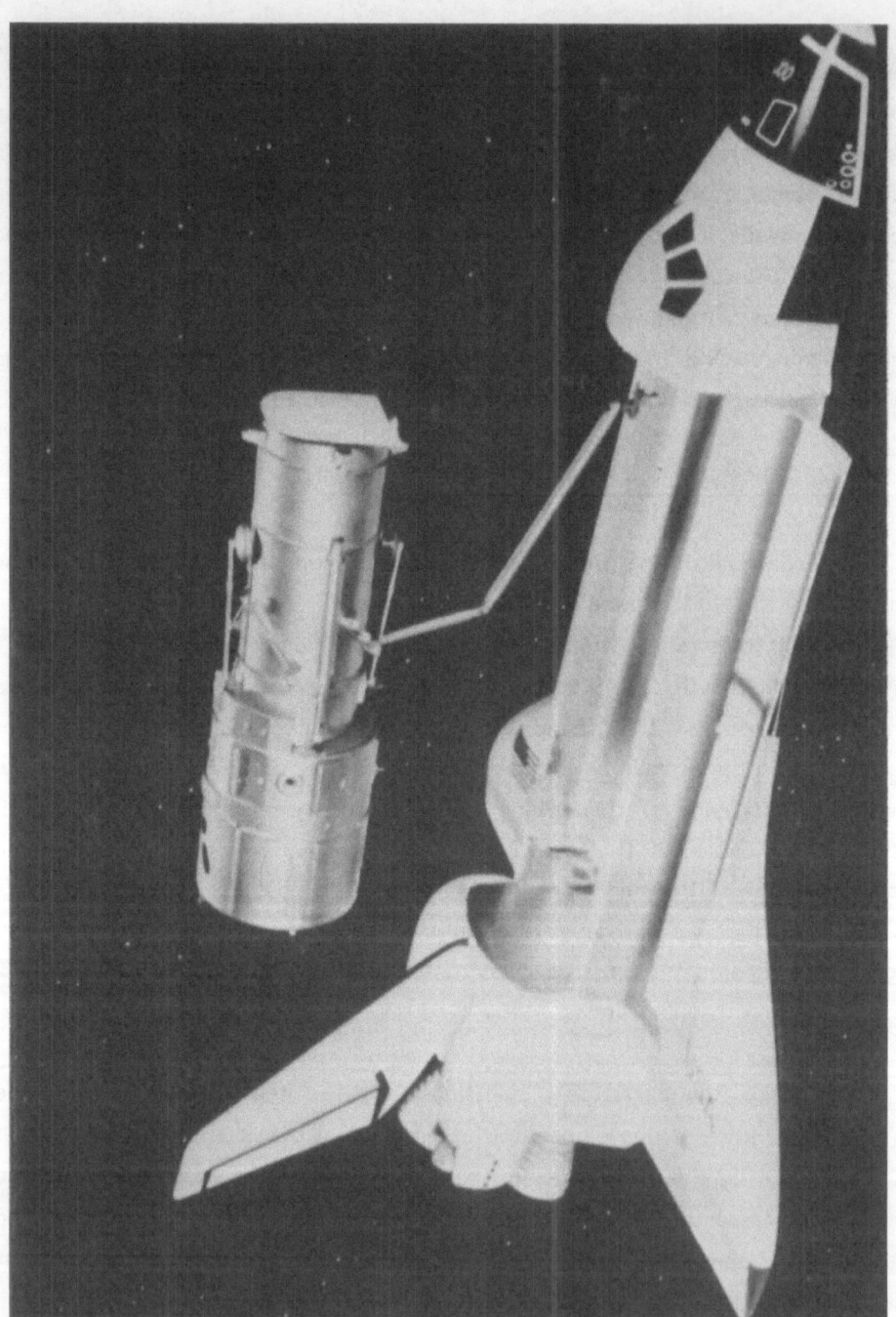

Figure 37 The NASA Space Telescope (Courtesy of NASA).

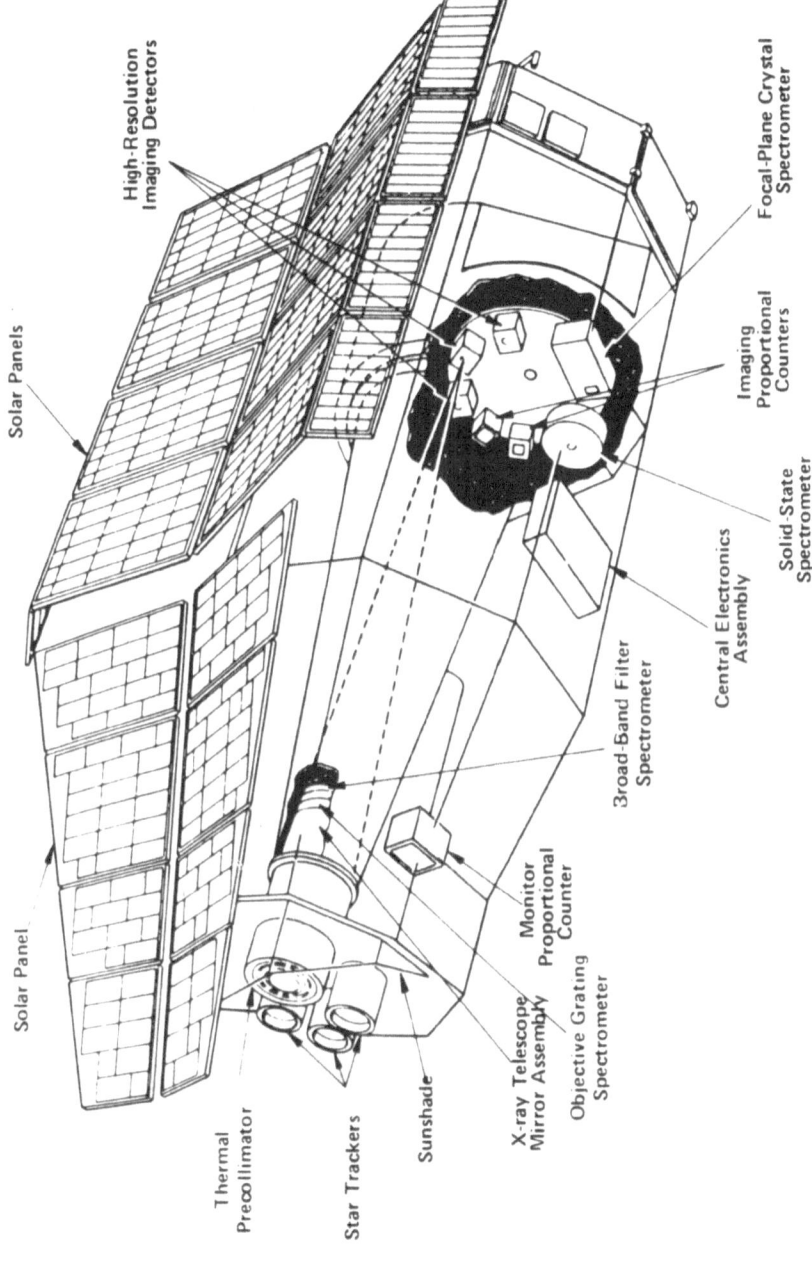

Figure 38   The Einstein X-ray Observatory (from N. Geril , 1981 , Perkin Elmer Technical News , 9 , 32) .

In addition to the imaging capability , there will be spectrographs and photometers which represent the best of modern astronomical technology . The fundamental scientific importance of this project is described in detail in a number of articles . A summary of my own views is included in the recent NASA publication NASA CP-2244 entitled 'The Space Telescope Observatory' which also contains authoritative articles on the telescope and all its instruments by the project leaders .

The Space Telescope will be at least a 15-year project and all astronomers will be able to apply for observing time; there are , however , some other special programmes which are complementary to the capabilities of Space Telescope. One particularly interesting project is that of a satellite devoted to spectroscopic studies of the short ultraviolet waveband from about $800\mathring{A}$ to $1200\mathring{A}$. The sensitivity of Space Telescope decreases very rapidly at wavelengths shorter than about $1200\mathring{A}$ because of the coatings of the mirrors . The special importance of this waveband is that many of the most important elements have strong emission and absorption lines in this waveband . Of particular significance are the absorption lines of deuterium which is not only of crucial importance cosmologically , but is also a tracer of the chemical evolution of the interstellar gas in our own and other galaxies . To give a simple example , if we find regions of the interstellar gas which are deficient in deuterium , this must mean that the gas in that region has all been processed through stars . On the other hand , if we find regions in which we know there has been a great deal of star formation and there is still a high deuterium abundance , we must assume the deuterium arrived there only recently . This is the type of argument which can be used to argue that our Galaxy is continuously accreting matter from intergalactic space .

A project of this type called **MAGELLAN** has been developed by the European Space Agency and is currently one of the candidate projects for selection as a future mission .

6.4 The X-ray Waveband

One of the great adventures of modern astronomy has been the development of X-ray astronomy . From the first observations in 1962 through the pioneering epoch of the **UHURU** satellite to the great success of the **Einstein** X-ray satellite Observatory , we have seen the subject grow from the first hints of a new discipline into one of the major research areas of modern astronomy .

The Einstein X-ray Observatory (Figure 38) , in particular , was the first complete X-ray observatory to be flown with a suite of instruments similar to that of a well equipped ground-based optical telescope . Some projects of special significance for cosmology have been the following :

(a)   the counts of X-ray sources to very faint X-ray intensities . These have shown that the quasars which are all strong X-ray sources exhibit strong evolutionary effects similar to those found in the optical and radio wavebands. These counts have enabled estimates to be made of the contribution of discrete X-ray sources to the observed X-ray background radiation (Figure 12) . It appears that active galaxies and quasars can account for a considerable part of the background , although exactly how much depends on more precise knowledge of the statistics of different classes of active galaxy .

(b)   Detailed studies have been made of the X-ray emission from clusters of galaxies . One of the predictions of some theories of the formation of clusters of galaxies was that , when they first form , they should be strong X-ray sources . This is because the large quantitites of gas which collapse to form a protocluster must dissipate the energy of collapse and the gas should be heated to such a temperature that it becomes an intense X-ray emitter. It turns out that this phenomenon has not been observed and the implications of this result for cluster formation and evolution are unclear . Obviously , we require a much more substantial data base to establish the typical X-ray properties of clusters at earlier epochs .

(c)   Everytime a new waveband becomes available for cosmological study , cosmologists investigate whether there exist any standard candles or rigid rods which might be used in the classical cosmological tests . At present we do not know if they exist and it requires a larger data base to establish whether or not there are new possibilities . There do exist potentially some physical mechanisms which could lead to the estimation of distances . For example , there exists a good physical reason why X-ray sources should not radiate much more than a certain X-ray luminosity . This is known as the Eddington limiting luminosity and amounts to the statement that if an X-ray source becomes too luminous , it will blow away by radiation pressure the basic source of energy which is the gravitational infall of matter . A number of the X-ray sources in our Galaxy approach or just exceed this limit . If it were possible to demonstrate that this limiting luminosity applies for X-ray sources in extragalactic systems , measurement of their apparent brightnesses could yield their distances by application of the inverse square law . Another exciting possibility of measuring distances is to use the X-ray luminosity of the intergalactic gas in clusters in conjunction with radio observations of the microwave background radiation in the same direction. Theoretically , it is expected that the radiation of the microwave background which passes through the gas will be influenced by scattering by the hot electrons in the gas . Where there is hot gas there should be a very small decrese in the intensity of the background to the long wavelength side of the peak of the spectrum of the microwave background radiation and an increase to the short wavelength side . The combination of the X-ray , radio and optical observations of the cluster end up providing sufficient data on the properties of the cluster to determine its physical size independent of its distance . The distance of the cluster r can than be found by comparing its angular size Θ with its physical size D , r = D/Θ . If this type of physical argument could be made to work , it would be of great importance in cosmology .

d)   We have already mentioned the problem of the origin of the X-ray background radiation . We expect at least part of this to be due to discrete X-ray sources but it is not at all clear whether they can account for all of it . A particular problem io tho origin of the rather sharp turn-over in the spectrum at about 40 keV (see Figure 12) . In any picture in which the background is due to the superposition of discrete sources , it is always difficult to produce any sharp features in the background radiation because of the intrinsic scatter in the properties of sources and because of the smearing effect of redshift . One possibility is that the background is due to the radiation of intergalactic gas at a high temperature . The main radiative process for gas at temperatures of about 10^7 - 10^9 K is known as bremsstrahlung and it has a characteristic spectrum which can explain rather precisely the X-ray background spectrum in the range 1-100 keV. The temperature of the gas would have to be about 300 million degrees. It is not at all clear at present if this interpretation is correct or where the gas , if it exists , is located. It might be a diffuse component of intergalactic material or it might be much more clumped in intergalactic clouds. A problem with this picture is the very large amounts of energy needed to heat so much gas to a high temperature. The precise origin of the X-ray background radiation is far from settled and deserves much further study by both observers and theorists.

The next important project in X-ray astronomy will be the **ESA EXOSAT** mission which will be launched in the coming year. Then there will be the German **ROSAT** programme in which a deep survey of the whole sky will be made with a telescope similar to the principal Einstein observatory telescope and a soft X-ray telescope provided by UK astronomers which will survey the sky in this relatively unknown waveband.

Perhaps the most exciting future project is the **NASA Advanced X-ray Astronomy Facility (AXAF)** (Figure 39) which will provide all astronomers with the same type of observing power in the X-ray waveband which the Space Telescope provides in the optical and ultraviolet wavebands. The four areas identified above can all be addressed very effectively by such an instrument as well as a host of related astrophysical problems of the origin and evolution of X-ray sources. It is projected that this facility will become available in the 1990s.

6.5 Gamma-Ray Wavelengths

Gamma-ray astronomy is among the most difficult of the space sciences because , in general , the fluxes of gamma-ray photons are very small and very long integrations are necessary to observe sources. Gamma-ray emission from the plane of our own Galaxy has been detected and this gives information about

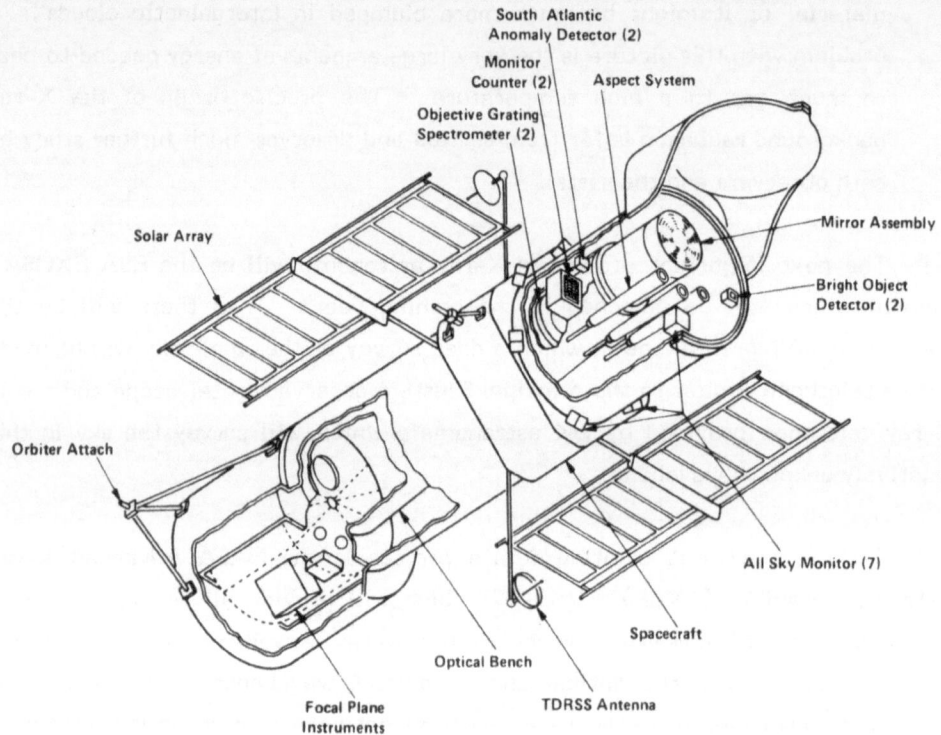

Figure 39   A schematic diagram showing a concept for the NASA Advanced
X-ray Astronomy Facility (from N. Geril , 1981).

Figure 40   A concept for an 8-metre optical-ultraviolet telescope in space
(from J.J. Russo and A.N. Brunner , 1981).

cosmic rays and cold gas in our Galaxy . About 12 gamma-ray sources belonging to our own Galaxy are known and only one certain identification with an extragalactic object has been made , that of the quasar 3C 273 . Even these results are major achievements in a very difficult field and are thanks to the efforts of the teams which built the **SAS-2** and **COS-B** satellites .

Therefore , the cosmological importance of studies of gamma-ray sources remains to be established by future generations of gamma-ray satellite . Among these the NASA **Gamma Ray Observatory (GRO)** is of special importance and should provide gamma-ray astronomers with a facility for making the first systematic observations of active galactic nuclei . There remains the question of the origin of the gamma-ray background . We know from studies in the hard X-ray waveband that some active galaxies have hard X-ray spectra which suggest that they will also be strong gamma-ray sources and it seems quite plausible that much of the gamma-ray background can be attributed to such sources . However this remains to be demonstrated on the basis of direct gamma-ray observations of the sources by future gamma-ray observatories .

6.6 Other Future Projects

The above list of projects is by no means complete but it highlights some projects of special importance for cosmology . In addition , many other exciting projects are being converted from gleams in the eye of the astronomer into feasible astronomical projects , the only limits being those of the imagination of the astronomer and the more important limits of funding and manpower. Examples of these types of project include a very large space telescope for optical and ultraviolet observations . An aperture of up to 10m (Figure 40) would represent a huge increase in scientific capability over even the Space Telescope. Space optical and infrared interferometry could provide very long baselines for a wide range of projects from measurements of the diameters of stars to the inner recesses of quasars and active nuclei . Very long baseline radio interferometry from space probes could provide ultrahigh resolution images of the rapidly expanding radio components in quasars and active nuclei . These types of project are probably for the next century rather than the present .

7 . The Ultimate Goal

The progress already made in astrophysical cosmology is quite staggering . We are now able to discuss seriously and astrophysically questions which would have been considered the purest fantasy 20 years ago . We have described a Universe in which we can explain many things using the best of modern physics. In many ways , these studies are only in their infancy and a very great deal of observation , interpretation and theory needs to be done to put flesh on the skeleton of the theory we have developed . There are two major areas which I can identify as major goals for future study . From the point of view of the astronomer , his task must be to confirm or disprove the rather elaborate picture of the physical evolution of the Universe which we have built up over the last 6 sections . The above scenarios must be subjected to the most searching scrutiny. It only needs one part of the jigsaw to be proved incorrect , for the whole picture to come into question . Thus , the main thrust of our observing programmes from the ground and from space must be the confrontation of the detailed physics of the hot big-bang with observational data .

The second aspect concerns the major problems which we have been building up throughout this exposition . We have already met two of them . They are :

1)   Why is the Universe asymmetric with respect to matter and antimatter?

2)   Where did the primordial fluctuations from which galaxies form come from?

The third problem is :

3)   Why is the Universe so uniform now on a large scale?

The origin of the third problem may be understood from the consideration of the "horizons" as discussed in Section 4 . There we showed that at any epoch in the Universe we can only communicate information a maximum distance of about ct where t is the age of the Universe . Thus as we go into the past , smaller and smaller regions of space can communicate with each other . In fact , when the Universe is about 1 minute old , only regions containing the mass of a single star can communicate at the speed of light . There is therefore a profound problem in knowing how it was that the Universe knew that it had to end up looking so isotropic at the present time . There is just not time for different bits of the Universe to reorganise themselves in such a way that they all look the same on average . According to the classical picture , this is yet another property which we would have to build into the initial conditions of the Universe .

We are therefore driven into the wholly unsatisfactory position of having to endow the initial conditions from which the Universe evolved with some highly specific properties . Specifically ,

a)   it must be <u>baryon asymmetric</u> at a level of about 1 part in 1,000 million;

b)   it must be <u>isotropic</u> from the beginning;

c)   it must contain <u>inhomogeneities</u> at a level of about one part in 10,000 .

If one begins with a completely arbitrary set of initial conditions for the Universe it is most unlikely that the Universe as we know it would come about .

It may well be that the answers to all these fundamental problems lie in some remarkable recent advances in the theory of elementary particles and the possibilities which have recently opened up of <u>Grand Unified Theories</u> of elementary particles . In addition , we know that at the very earliest epochs , $t < 10^{-43}$ second , our theory of gravitation is incomplete and should be quantised .

These theories offer us some new ways of tackling these problems . First of all , in the Grand Unified Theories , the Universe is symmetric at the very highest temperatures . When the Universe "cools down" through a critical temperature , "symmetry breaking" occurs which results in one type of particle being favoured over another . This process has been compared with the phenomenon of ferromagnetism whereby , above a certain temperature , the Curie temperature , there is no preferred magnetisation state for materials like iron . However , when the temperature drops below the Curie temperature , there is a spontaneous symmetry breaking which results in a preferred orientation of the magnetic domains . In the same way , in the early Universe , there could have been symmetry between matter and antimatter which is broken when the Universe cools below a certain critical temperature . According to particle theorists , this process of symmetry-breaking could produce the baryon asymmetry of our Universe . Second , there are processes by which fluctuations of finite amplitude can be generated in the very early Universe . These involve quantum fluctuations at the earliest epochs or processes which take place at the various phase transitions in the early Universe .   Third , there exist means by which the Universe can be smoothed out on scales far beyond the horizon scale . Some of these involve the quantum nature of the very early Universe and others involve strange changes which may take place in the equation of state of matter at extremely high temperatures .

This is not the place to go into the details of these scenarios and much of it is still highly speculative . The important point is that questions which until recently were profound obstacles in building a convincing model for the hot big-bang may now be capable of being resolved by advances in particle physics and gravitational theory . What we may be witnessing is a unique coalescence of the apparently disparate sciences of cosmology and elementary particle physics into a unified picture of the origin of our Universe . At last , a real answer may be given to questions of the origin of the Universe itself . The answers stretch back to the quantum area of cosmology when time itself ceases to mean anything .

We have come a long way from the practicalities of the real Universe to the epochs when time itself no longer exists . I would emphasise , however , that we are dealing with a house of cards . If we remove one card , the whole structure may topple and we must start again from scratch . This is the excitement of astrophysical cosmology . Our models of the Universe live dangerously and are subject to constant assault by observation and theory . In this endeavour , I hope I have convinced you that Space Science will play an essential and growing role in what may turn out to be the greatest intellectual achievement of all time.

REFERENCES AND FURTHER READING

The sources of material described in this review are so vast that I have made no attempt to reference every statement . Details of most of the topics can be found in the following selection of books and reviews .

G.O. Abell and P.J.E. Peebles (eds), 1980 . "Objects of High Redshift", IAU Symposium No. 92 , D. Reidel Publishing Company .

H.A. Brück , G. Coyne and M.S. Longair (eds), 1982 . "Astrophysical Cosmology" , Proceedings of the Vatican Study Week , Publications of Pontifical Academy of Sciences .

A.N. Bunner , et al. , 1981 . "Space Astronomy" , Perkin Elmer Technical News , $\underline{9}$, No. 1 .

S.M. Fall and D. Lynden-Bell (eds), 1981 . "The Structure and Evolution of Normal Galaxies" , Nato Advanced Study Institute , Cambridge University Press .

J.E. Gunn , M.S. Longair and M.J. Rees , 1978 . "Observational Cosmology" , 8th Advanced Course of Swiss Society of Astronomy and Astrophysics , Geneva Observatory .

C. Hazard and S. Mitton (eds) , 1979 . "Active Galactic Nuclei" , Nato Advanced Study Institute , Cambridge University Press .

D.S. Heeschen and C.M. Wade (eds) , 1979 . "Extragalactic Radio Sources" , IAU Symposium No. 97 , D. Reidel Publishing Company .

M.S. Longair , 1971 . Reports on Progress in Physics , $\underline{34}$, 1125 .

M.S. Longair and J. Einasto (eds) , 1978 . "The Large Scale Structure of the Universe" , IAU Symposium No. 79 , D. Reidel Publishing Company .

M.S. Longair and J. Warner (eds), 1979. "Scientific Research with the Space Telescope", IAU Colloquium No. 54, NASA Special Publication CP-2111.

P.J.E. Peebles, 1971. "Physical Cosmology", Princeton University Press.

P.J.E. Peebles, 1980. "The Large Scale Structure of the Universe", Princeton University Press.

M. Rowan-Robinson, 1981. "Cosmology", Clarendon Press, Oxford.

E. Schatzman (ed), 1973. "Cosmology - Cargese Lectures on Theoretical Physics", 6, Gordon and Breach.

D.W. Sciama, 1971. "Modern Cosmology", Cambridge University Press.

S. Weinberg, 1972. "Gravitation and Cosmology", John Wiley and Sons.

Ya. B. Zeldovich and I.D. Novikov, 1982. "The Structure and Evolution of the Universe", University of Chicago Press (in press).

The reader is also referred to recent issues of "Annual Reviews of Astronomy and Astrophysics" for up-to-date authoritative surveys of many of the topics referred to in this review.

CHAPTER VI

CONCLUDING REMARKS

by Professor A.G. Massevich

Astronomical Council , USSR Academy of Sciences ,
Pyatnitskaya ul. 48
109017 , Moscow , USSR

This will be a very brief conclusion . We have been listening to five brilliant lectures , covering a very vast field of astronomical research , starting from the Sun . Professor Pecker , in his talk about the physics of the Sun , included stellar seismology and neutrino astronomy . Professor Bonnet gave us a beautiful picture of the solar-terrestrial relations based on the newest research and results obtained from instrumentation on-board satellites , especially concerning the structure of the magnetosphere . In the lecture of Professor Hanbury Brown we heard about the stellar world and about a very sophisticated instrumentation which allows to measure diameters , shape , and flattening of far-away stars . Professor Lewin in his lecture gave us a very interesting view of the newly-discovered celestial bodies , which could have been discovered only by means of space science and space technology . And Professor Longair introduced us to the vast Universe . I don't think that I have to make a summary of each of these reports because they were very comprehensive , very clear and very beautifully presented and illustrated by many slides and view-graphs .

Astronomers have been dreaming about observations beyond the atmosphere as long as astronomy exists . Henry Norris Russell wrote in one of his books that all good astronomers , after death , instead of going to paradise would be transferred to the Moon . He wanted to stress the importance of being able to observe beyond the atmosphere . Fortunately , astronomers now do not have to be dead in order to observe beyond the atmosphere ! Now they can in good health conduct their observations by automatic spaceships and also by manned spacecraft .

R. M. West (ed.), Understanding the Universe, 227–230.
© *1983 by D. Reidel Publishing Company.*

Professor Bonnet showed us today the first historical extra-terrestrial spectra of the Sun taken as early as in 1946 . I am sure many of you remember the very important paper that Professor Lyman Spitzer published , also in 1946 , with the title "The Astronomical Advantage of Extra-terrestrial Observations" . In the past twenty-five years since the first satellites allowed us to go beyond the atmosphere , astronomy has probably gained more than any other science from space technology . Astronomers really deserved it , because for more than 300 years after the first telescope was built by Galileo , they dealt only with optical light and by analysing it they had to discover the surrounding world . Only in the last fifty years an additional window in the radio range was opened to help them understand what happens in the Universe .

Space research has now given us , besides what has already been said in the lectures , a clearer picture of our own Earth and of the origin of our solar system . The discovery of the X-Rays , X-Ray sources and identification of them with the neutron stars in close binary systems has given the missing links in order to construct a comprehensive evolutionary scenario , for massive stars as well as for stars of small masses . And , of course , the number of examples could be increased , showing which new possibilities have become available with space technology and space research .

What was very important and very interesting for me , and I think for many of those present here also , was that in all of today's lectures the necessity and importance of ground-based astronomy and extra-terrestrial astronomy working hand in hand was directly or indirectly stressed . This is really very important because in the beginning of the space era there was a slight tendency that perhaps extra-terrestrial astronomy would make ground-based astronomy superfluous . Even some publications appeared in which it was said that there was no need for construction of large telescopes because the space telescopes could cover all this . But we have seen during these twenty-five years that the necessity for ground-based observations is very large . A very good example , I think , was given today by Professor Lewin when he told us that one of the gamma bursters has possibly been identified on a photographical plate obtained as early as 1928 and that , also by a ground-based telescope , the possible optical counterpart has now been found as a very faint star . I think there was a similar example when the variability of one of the quasars was first discovered also by studying an old collection of plates for variable stars at the Sternberg Institute in Moscow . Again in this field a lot of examples could be given , so I think that ground astronomy has a large , important future .

Professor Hanbury Brown stressed today that it is very important to have new instrumentation which will give us new and very important results in space and on the ground . In a conference like UNISPACE 82 , when we talk mostly about the possibilities for developing countries to share the benefits of space technology , it is very important to stress (as also mentioned in the draft report) that we must share fully the results from space science .

In every country participating in this field , a group of scientists should be established , consisting of specialists in basic science who can plan and develop the application of space science in their country . In the recommendations it is particularly stressed that astronomy and basic sciences should be introduced in the universities of these developing countries , because experience has shown that people with a scientific background were the first to be deeply engaged and to make progress in space applications .

In the name of the UNISPACE 82 Secretariat I would like to thank very much the speakers who have come today at the invitation of the Executive Committee of the International Astronomical Union . I am sorry that we didn't have a very large audience because the lectures given really deserved to be heard by many more . Unfortunately , the planning of the UNISPACE 82 conference was so that at the same time several parallel meetings are going on , and we could do nothing to improve this .

The Assistant General Secretary of the IAU , Dr. Richard West , asked me to tell you that the lectures will be published by Reidel Publishing Company as a separate book . My special thanks are due to Professor Strömgren who has been so kind to come here and to chair our session and , as Professor Pecker already said , I can only repeat that all the present astronomers consider him as their spiritual

teacher . So thank you very much for coming here and attending the UN/IAU
International Astronomy Seminar today .

UNISPACE 82

The Second United Nations Conference on the Exploration and Peaceful Uses of Outer Space , or UNISPACE 82 , took place in Vienna , Austria , 9 to 21 August 1982 . It brought together more than one thousand responsible politicians , scientists , technicians and industrials from about 100 countries and international organisations to discuss Space Sciences and Techniques . One of the principal goals was to discuss the impact of these on the socio-economical development of mankind .

The agenda of UNISPACE 82 had three large headings : Summary and Future Perspective of Space Science and Techniques , Applications of Space Science and Techniques , and International Cooperation and the Role of the United Nations .

The extensive preparations for UNISPACE 82 commenced already in 1978 . The Secretary-General of the Conference was Professor Yash Pal (India) ; the Deputy Secretaries-General were Professor Jerry Gray (USA) and Professor Alla Massevich (USSR) .

The Conference provided a unique opportunity for the international community to gather together to consider in great detail the complex issues of this new global concern . It dealt with the entire gamut of space sciences , technologies and applications from scientific , technical , political , economic , social and organisational points of view . Although legal issues were not on the official agenda , the Conference also discussed the legal implications of issues on the agenda . Similarly , though the agenda was limited to the peaceful uses of outer space , the Conference discussed at length , particularly during the general debate , the growing international concern relating to military activities in outer space .

Acting by consensus , the Conference adopted its report , which has been published by the United Nations . The General Assembly of the United Nations at its 37th session adopted a resolution urging all Member States and concerned international organisations to implement the recommendations of the UNISPACE 82 Conference .

The International Astronomical Union

The International Astronomical Union was founded in 1919 . It is an international, non-governmental organisation, supported by Adhering Organisations (mainly Academies of Science) in 49 countries . The Union is also an association of individual members, currently numbering more than 5,200 . It was founded for the purpose of developing astronomy in all its aspects, through international cooperation, and for safeguarding of the interests of astronomy, especially in respect of international agreement . The IAU is a founding member of the International Council of Scientific Unions (ICSU) .

The IAU is very diverse in character and its scientific activity is reflected in the work of its 40 Commissions, which refers to all phenomena in outer space . The IAU also organises meetings at regular intervals in the member countries and holds triennial General Assemblies with the participation of several thousand members for the discussion of current research and for the planning of future projects, frequently involving wide international collaboration .

Special attention is paid to the development of astronomy in countries where this science is not yet well established, e.g. by means of Schools for Young Astronomers (in cooperation with UNESCO), a Visiting Lecturers' Programme, support for travels, mainly by Young Astronomers to major astronomical centres for extended study, the compilation of education materials, etc.

Further information about IAU can be obtained from the IAU Secretariat, 61, avenue de l'Observatoire, F-75014 Paris, France .

About the Chairman

Professor emeritus Bengt G. Strömgren obtained his PhD in astronomy from the Copenhagen University in 1929 and became Professor at the Copenhagen University in 1938 . After a long stay in the United States where he was Professor at the University of Chicago , Director of the Yerkes and McDonald Observatories , and later Professor at the Institute for Advanced Study in Princeton , he returned to Copenhagen in 1967 as Professor at the University and at NORDITA until his retirement in 1978 . Professor Strömgren was the General Secretary (1948-1952) and President (1970-1973) of the International Astronomical Union .

Professor Strömgren's research interests cover many fields including the study of the detailed structure of the Milky Way , our galaxy .

About the Speakers

Professor Jean-Claude Pecker entre au CNRS en 1946 à l'Institut d'Astrophysique comme Attaché de Recherches et prépare sa thèse qui porte essentiellement sur les différents aspects de la théorie des atmosphères liées à la notion de type spectral . Il devient ensuite Chargé puis Maître de Recherches au CNRS , toujours à l'Institut d'Astrophysique . Il crée le Service d'Astrophysique Générale à l'Observatoire de Meudon et en 1963 il est élu au Collège de France à la Chaire d'Astrophysique Théorique créée pour lui . Il assume la fonction de Directeur de l'Observatoire de Nice en 1962 ; en 1971 il prend la direction de l'Institut d'Astrophysique . Il a été Secrétaire Général de l'Union Astronomique Internationale (UAI) ; il a présidé le Comité des Sciences de la Commission Nationale pour l'Unesco , ainsi que diverses commissions de l'UAI , des organismes internationaux comme l'ESO ou l'ESRO (devenu ASE) .

Les activités principales de Jean-Claude Pecker dans le domaine de l'astronomie ont concerné la théorie des atmosphères stellaires, la physique solaire, la physique des relations Soleil-Terre, les composantes poussièreuses du milieu interstellaire, enfin les cosmologies alternatives que suscite l'observation de décalages anormaux vers le rouge.

Jean-Claude Pecker a écrit plus de 400 articles sur les différents domaines de l'astronomie, destinés à des spécialistes, une centaine d'articles destinés à un public plus étendu, ainsi qu'une dizaine d'ouvrages.

En dehors de ses activités purement astronomiques, Jean-Claude Pecker s'est intéressé à un assez grande nombre de problèmes liés notamment à la défense des droits de l'homme et à la lutte pour une information scientifique réelle, donc à la lutte contre les pseudo-sciences.

Professor Roger M. Bonnet obtained a PhD from the Paris University in 1961. Since then he has worked at the Centre National de la Recherche Scientifique where he was appointed Directeur de Recherches in 1977. He has also been Director of Laboratoire de Physique Stellaire & Planétaire since 1969. His scientific activity has mainly been aimed at the study of the ultraviolet spectrum of the sun by means of balloons, rockets, and artificial satellites. He is now involved in several space experiments, including the ESA GIOTTO mission, and the Soviet VEGA programme. From 1983, he is Scientific Director of ESA.

In 1936 **Professor R. Hanbury Brown** joined the original team which developed radar in the U.K. From 1949 he worked at the Jodrell Bank Experimental Station on Radio Astronomy and was appointed Professor of Radio Astronomy in 1960 at the University of Manchester. He developed the stellar intensity interferometer at Narrabri in Australia, and was appointed Professor of Physics (Astronomy) at the University of Sydney, from where he retired in 1981. Professor Hanbury Brown is the President (1982-1985) of the International Astronomical Union. His astronomical research is mainly directed towards the measurement of the fundamental parameters of stars.

Professor Walter H.L. Lewin obtained his PhD in Physics in the Netherlands . He later went to the United States and is currently Professor at the Massachusetts Institute of Technology , Cambridge .   He is one of the pioneers in X-ray Astronomy and has been closely involved in research with earth-orbiting satellites since 1966 .

Professor Malcolm S. Longair obtained a PhD in Radio Astronomy in 1967 at the Cavendish Laboratory . He worked at this laboratory for some time and , after a period in the Soviet Union on a Royal Society Exchange Fellowship , he was Visiting Professor at the California Institute of Technology and the Institute for Advanced Study , Princeton .   Since 1980 he has been Astronomer Royal for Scotland at the Royal Observatory , Edinburgh . His principal research interests are high energy astrophysics and cosmology .

Professor Alla G. Massevitch was Deputy Secretary-General of UNISPACE 82 and is Vice President of the Astronomical Council of the USSR Academy of Sciences . She is Professor of Astrophysics at Moscow University and Chairman of the Intercosmos Program for Satellite Geodesy and Geodynamics . She is in charge of all astronomical observations of artificial satellites in the USSR .   Her main research interests are in the fields of satellite geodesy and the evolution of stars .

Astronomical and Physical Constants

This list is not intended to be exhaustive , but merely to help the reader who is not accustomed to "astronomical" units used in this book .

$k = 10^3$ (kilo) ; $M = 10^6$ (Mega , million) ; $G = 10^9$ (Giga , billion)

1. Distance

1 A.U. (Astronomical Unit) =	1.496×10^8 km
1 pc (parsec) =	3.086×10^{13} km
1 light year =	9.461×10^{12} km
1 μm (micrometer) =	10^{-6} m
1 nm (nanometer) =	10^{-9} m
1 A (Angstrom) =	10^{-10} m

One Astronomical Unit is the mean distance from the Sun to the Earth , i.e. the mean radius of the Earth's orbit around the Sun . One parsec is the distance from which this radius (1 A.U.) subtends an angle of 1 arcsecond . One light year is the distance travelled by light in one year (1 pc = 3.26 light years) . The Angstrom is used in spectral work .

2. Mass

$1 M_\odot$ (solar mass) $= 1.989 \times 10^{30}$ kg

3. Velocity

c (velocity of light) $= 2.998 \times 10^8$ ms^{-1}

The redshift z of an astronomical object measures the velocity along the line of sight , e.g. $z = 1$ corresponds to 180.000 kms^{-1} (including the relativistic effect) .

4. Energy

1 erg $=$ 10^{-7} J (joule)

1 eV (electron volt) $=$ 1.60 $\times$ 10^{-19} J

Light quanta (photons) carry energy that is related to their wavelength λ by $E = h\nu$, where h is Planck's constant and ν is the frequency ($= 1/\lambda$). The frequency is measured in Hz (Hertz) ; 1 Hz = 1 cycle s^{-1} . For instance :

	Wavelength (λ)	Frequency (ν)	Energy
Radio waves	1 m	3×10^8 Hz	1.2×10^{-6} eV
Micro waves	1 mm	3×10^{11} Hz	1.2×10^{-3} eV
Infrared radiation	10 m	3×10^{13} Hz	0.12 eV
Visible	5000 A	6×10^{14} Hz	2.5 eV
Ultraviolet	1000 A	3×10^{15} Hz	12.4 eV
X-ray	1 A	3×10^{18} HZ	12.4 KeV
Gamma-ray	10^{-12} m	3×10^{20} Hz	1.24 MeV

5. Power (Luminosity)

1 W (watt) = 1 J s^{-1}

1 $L_\odot$ (solar luminosity) = 3.9 $\times$ 10^{26} W

6. Flux (Brightness)

The brightness of astronomical objects is often measured in **magnitudes** . From ancient times , magnitude 1 corresponds to the brightest stars and magnitude 6 to those which can barely be seen with the unaided (naked) eye . A revision of this **logarithmic** scale in the 19th century fixed the interval between two magnitudes as 2.512 . Thus an object of magnitude 3 is 2.512 times brighter than one of magnitude 4 and exactly 100 (= 2.512^5) times brighter than one of magnitude 8 . The faintest objects that can now be observed with ground-based telescopes have magnitude 25 ; the Space Telescope (cf. p. 220) may reach magnitude 29 .Bright stars , planets , the Sun and the Moon have negative magnitudes .

A **colour index** of an object is the magnitude difference , as measured in different spectral regions , e.g. (B-V) is the difference of brightness in blue ($\sim$4500 A) and green-yellow ($\sim$5500 A) light . The colour index is mainly a measure of the temperature .

The magnitude of an object as measured on the Earth is called **apparent magnitude** . To compare the intrinsic brightness of different objects , the **absolute magnitude** is calculated , i.e. the apparent magnitude the object would have if it were situated at a distance of 10 parsec . E.g. the Sun's apparent magnitude is -26^{m} , but its absolute magnitude is +5^{m} .

The apparent brightness of an object falls of at the **square of the distance** , i.e. it would be 4 times fainter if it were at twice the distance .

7. Temperature

K (degrees Kelvin) = temperature above absolute zero (-273°C)

Astrophysical temperatures range from near the absolute zero in interstellar space to several million K in the interior of stars . Even higher temperatures may be reached at supernova explosions and , of course , during the first moments of the Big Bang .

SUBJECT INDEX